"Innovative, provocative, and essential reading, for all those concerned about the state of the world and the purposes of education. Read it, be disturbed and challenged. Question your basic premises about education. Renew the vision. Take committed action."—Professor David Hicks, Bath Spa University College, UK

"Helps us address the most important educational need of our age: how to reorient our education systems so that they guide us towards a sustainable future. A reading list must for everyone involved in education—from Ministers right up through the system to teachers and parents."—John Fien, Director, Griffith University Ecocentre, Australia

Stephen Sterling is an independent consultant working in the academic and NGO fields in the UK and internationally. He was involved in developing the MSc in Environmental and Development Education at South Bank University, London, where he is an academic tutor. He has an extensive publications record, including *Good Earth-Keeping: Education, Training and Awareness for a Sustainable Future* (UNEP UK 1992), *Education for Sustainability* (Earthscan 1996), and *Education for Sustainable Development in the Schools Sector* (Sustainable Development Education Panel 1988). He has been a long-term consultant with WWF-UK on their professional development programme. He is a member of the IUCN Commission on Education and Communication, of the UNED Forum Education Task Group, and the Oxfam Advisory Group on Global Citizenship, and is an external examiner for Bath Spa University College. He is currently completing a doctoral thesis on systems thinking, education and sustainability.

Schumacher Briefing No. 6

SUSTAINABLE EDUCATION

Re-visioning Learning and Change

Stephen Sterling

published by Green Books
for The Schumacher Society

First published in 2001
by Green Books Ltd
Foxhole, Dartington, Totnes,
Devon TQ9 6EB
www.greenbooks.co.uk
greenbooks@gn.apc.org

for The Schumacher Society
The CREATE Centre, Smeaton Road,
Bristol BS1 6XN
www.schumacher.org.uk
schumacher@gn.apc.org

Cover design by Rick Lawrence

Printed by J.W. Arrowsmith Ltd
Bristol, UK

A catalogue record for this publication
is available from the British Library

ISBN 1 870098 99 4

CONTENTS

Acknowledgements

I would like to convey my sincere thanks to Herbert Girardet, David Hicks, John Parry, Jane Reed, William Scott and Martin Sterling, all of whom commented on drafts of the text; to John Fien, Morag Gamble, Siv Sellin, Gillian Symons and Caroline Walker, all of whom contributed material; and particularly to the editor, Sophie Poklewski Koziell. Special thanks to my wife Deborah for her support.

Foreword

by David Orr

If education is the solution, what is the problem? For many, the answer is transparently obvious: to prepare young people for careers in the global economy, working for one corporate behemoth or another. This is certainly true of the conversation about education presently taking place in the United States, which is long on performance standards and testing and short on how to encourage critical thinking, creativity, and ecological awareness. Why is this so? Part of the answer, I believe, is found in the progressive diminution of the idea of learning throughout the 20th century. Far removed from the tradition of the great philosophers, the discourse on education has become a technical subject requiring only efficient administration by technocrats. From this perspective the humanities and arts are expendable, because the goal of education is little more than to equip the young for the new information economy. The 'solution' is to fill every nook and cranny of the school with computers. The naiveté of this view is quite touching and just as wrong-headed. But this whittled-down version of education is also convenient to those whose interests are well served by a docile, but technically competent, public, otherwise unable to think critically or to act as citizens.

The stakes are too high, however, to let matters rest here. The world is coming apart at the seams. Only those caught deep in denial can ignore the mountain of scientific data about destructive effects of human activities on the Earth. Only those who choose not to see can ignore the human tragedies unfolding before our eyes evident in poverty, malnutrition, and violence. But only those equipped to discern and think critically will understand both the magnitude of the problems and the choices to be made if we are to create a decent and humane future. The disorder we see all around us reflects a prior disorder grounded in the paradigm of human domination that has now nearly conquered the entire world. That paradigm must be replaced by one that places us in the web of life as citizens of the biotic community. We must come to see ourselves as implicated in the world, not simply isolated, self-maximizing individuals. This battle will be won or lost in the schools, colleges, and universities around the world.

The challenge that Stephen Sterling sets before us begins with the assumption that "the qualities, depth and extent of learning that takes place globally in the next ten to twenty years" are critical for the human future. Most education simply reinforces practices and pathologies that cannot and should not be sustained over the long term. In contrast, he proposes an education that sustains the "whole person—spirit, heart, head and hands" and reclaims the authentic tradition of education to equip the young for lives of thought and purpose. This is a large idea of education that extends well beyond formal schooling toward "a vision of continuous re-creation or co-evolution where both education and society are engaged in a relationship of mutual transformation".

The upshot is that we must take education back from those who intend it to be centralized, homogenized, standardized, technologized, and industrialized. This is a struggle that we must win. But what might it mean to win such a struggle? I've read no clearer or more concise description than that given here by Stephen Sterling. Authentic education, as Sterling argues, has always been rooted in place and tradition. It is oriented to the pace at which the young learn well, not by imposed standards and schedules. It has always been open and participatory. Authentic learning engages, educes, encourages, and enthuses. Instead of rote learning, real education encourages spontaneity, insight, and reflection. Its aim has always been whole persons who are capable of thinking critically and living with compassion, energy, and high purpose.

The stakes have never been higher. We must, in short order, be about what Thomas Berry calls "The Great Work" of remaking the human presence on the Earth—how we provision ourselves with food, energy, materials, water, livelihood, health, and shelter. We must build authentic and vibrant communities that sustain us ecologically and spiritually. For this challenge we need a generation equipped to respond with energy, moral stamina, enthusiasm, and ecological competence. This requires a new understanding of ourselves and our place in nature and in time. This is *the* challenge of education.

David W. Orr is Professor, Environmental Studies Program Oberlin College, Ohio, USA. He is the author of *The Nature of Design* (Oxford University Press, 2001) and *Earth in Mind* (Island Press, 1994).

Aims of the Briefing

This Briefing sets out to:

• critique the prevailing educational paradigm from an ecological perspective;

• outline an alternative educational paradigm: one which helps sustain the 'whole person', communities, and the environment;

• outline the ecological basis of this framework and its implications for learning;

• present brief case studies of educational initiatives, institutions and ideas which are putting 'sustainable education' into practice;

• suggest design strategies and measures for policy-makers and practitioners that would help realize 'sustainable education'; and

• stimulate thinking and action on education that takes us closer to a sustainable future.

Summary of the Argument

1. The difference between a sustainable or a chaotic future is learning.
2. Most learning, however, is functional or informational learning, which is oriented towards socialization and vocational goals that take no account of the challenge of sustainability.
3. This has been reinforced in Western educational systems by the introduction of a managerial and instrumental view of education—which has paralleled economic restructuring in recent years.
4. This modernist educational paradigm derives from a broader social and cultural paradigm, which is fundamentally mechanistic and reductionist.
5. There is a poor fit between this dominant paradigm and our experience of increasing complexity, interdependence, and systems breakdown in our lives and the world.

6. Calls to assert 'education for sustainable development' or 'environmental education' within the framework of a mechanistic education paradigm can only meet with limited success, as such forms of education for change are marginalized and accommodated by the mainstream.
7. The real need is to change from *transmissive* towards *transformative* learning, but this in turn requires a transformed educational paradigm. Paradigm change is itself a transformative learning process.
8. Educators for change need a clearer understanding of the nature of an emergent ecological, participatory worldview from which a strong ecological educational paradigm and culture can be developed, whilst relating to the best humanistic education traditions.
9. The ecological, participatory worldview can be interpreted in terms of 'whole systems thinking'. This approach helps clarify the difference between a mechanistic and an ecological approach.
10. Whole systems thinking *extends, connects* and *integrates* the three aspects of paradigm: *ethos, eidos,* and *praxis* to reflect wholeness.
11. There is a parallel between the *social response* and the *educational response* to the challenge of sustainability, both of which may show progressive *levels of learning* as follows:

- no response (ignorance/denial/no learning);
- accommodatory response (adaptive learning, paradigm unchanged);
- reformatory response (reflective adaptive learning, paradigm modified);
- transformative response (critical and creative learning, changing paradigm).

12. Realization of a sustainable education paradigm requires *vision, image, design,* and action—at all levels—from all concerned with achieving healthy societies and ecologically sustainable lifestyles.
13. This is relevant to the whole of education and learning.
14. Time is critically short to make the change in education necessary to assure a secure future.

Introduction

"No problem can be solved from the same consciousness that created it. We have to learn to see the world anew."—Einstein

Seeing education anew

The key to creating a more sustainable and peaceable world is learning. It is the change of mind on which change towards sustainability depends; the difference of thinking that stands between a sustainable or chaotic future. The qualities, depth and extent of learning that takes place globally in the next ten to twenty years will determine which path is taken: either moving towards or further away from ecological sustainability.

This Schumacher Briefing is concerned with the re-visioning and re-orientation of education, particularly within the contexts of Western and Westernized cultures, given the urgent need for mass social learning 'towards sustainability'. Ever since the UN Conference on the Human Environment of 1972, there have been numerous high-level calls to address the most pressing issues of our age through realigned forms of education. But nearly thirty years later, and on the verge of the second Earth Summit in 2002, most education contributes daily to unsustainability, partly by default. At the same time, it does little to sustain the 'whole person'—spirit, heart, head and hands.

Back in 1973, E.F. Schumacher suggested that education was "our greatest resource" but also warned that unless it clarified "our central convictions" it would ultimately be a destructive force.[1] Yet in the last fifteen years or so, education has been re-structured and repackaged to conform to the philosophy and perceived needs of the market, and the managerial influence may now be seen in most Western and Westernized education systems across the world. This move towards the 'modernization' and globalization of education,

Box 1: Unsustainability and sustainability

Unsustainability . . .
"Between 1970 and 1995, the world lost 30% of its natural wealth, based on measures of the state of forest, freshwater and marine environments."[2]

"More than 113 million children have no access to primary education, 880 million adults are illiterate, gender discrimination continues to permeate education systems, and the quality of learning and the acquisition of human values and skills fall far short of the aspirations and needs of individuals and societies. . . . Without accelerated progress towards education for all, national and internationally agreed targets for poverty reduction will be missed, and inequalities between countries and within societies will widen."[3]

Sustainability . . .
"A sustainable society is one that can persist over generations, one that is far-seeing enough, flexible enough, and wise enough not to undermine either its physical or social systems of support."[4]

"The sustainability transition is the process of coming to terms with sustainability in all its deeply rich ecological, social, ethical and economic dimensions. . . . It is about new ways of knowing, of being differently human in a threatened but cooperating world . . . "[5]

for all its belief in being at the forefront of change, is in many senses behind the times. It is:

- still informed by a fundamentally mechanistic view of the world, and hence of learning;
- largely ignorant of the sustainability issues that will increasingly affect all aspects of people's lives as the century progresses;
- blind to the rise of ecological thinking which seeks to foster a more integrative awareness of the needs of people and the environment.

It is an example of what Laszlo calls the attempt "to cope with the conditions of the 21st century with the thinking and practices of the 20th". [6] In short, there is a very poor fit.

Our paradoxical times—of both great danger and opportunity, rapid change and a search for grounding and identity—require new vision in education. In contrast to the quasi-market model of education, we need instead a more intelligent, subtle, whole view of learning and education. One that builds from humanistic educational approaches in the past, but also takes full account of new developments relating to complexity theory, systems theory, learning theory and the pressing imperative of sustainability. One that values and sustains people *and* nature, that recognizes their profound interdependence. Such a view is more holistic, participative, and practical than the narrowly instrumental view that now dominates: in short, it is an *ecological* or relational view of education and learning. And it is a reflexive and postmodern view rather than a modernist view. This more holistic view, I believe, can become the next educational paradigm—but we need to articulate and develop it now if it is to emerge from the margins of educational thought, policy and practice.

This Briefing attempts to outline a democratic and ecological alternative to the dominant discourse which, in recent years and paralleling changes in the economic sphere, has swept all before it as if no alternative were possible. The title of the Briefing is 'sustainable education', to suggest the shift of educational culture that is required. Words have power. This is clearly demonstrated in the world of education, where managerialist language has almost replaced more traditional educational terminology and led to a narrowed discourse and practice. If we want a more humanistic, democratic and ecological educational paradigm, then we must find the ideas and language to help create it. The idea of 'sustainable education' is a powerful start.

The term 'sustainable education' implies whole paradigm change, one which asserts both humanistic and ecological values. By contrast, any 'education *for something*', however worthy, such as for 'the environment', or 'citizenship' tends to become both accommodated and marginalized by the mainstream. So while 'education for sustainable development' has in recent years won a small niche, the overall educational paradigm otherwise remains unchanged. Within this paradigm, most mainstream education *sustains unsustainability*—through uncritically reproducing norms, by fragmenting understanding, by sieving winners and losers, by recognizing only a narrow part of the spectrum of human ability and need, by an inability to explore alternatives, by rewarding dependency and conformity, and by servicing the consumerist machine. In response, we need to *reclaim* an authentic education which recognizes the best of past thinking and practice, but also to *re-vision* education and learning to help assure the future.

Deep learning and change

The possibility of a 'new educational paradigm' is based upon a very important distinction, between 'first order' change and 'second order' change, or between *first order learning* and *second order learning*. First order change and learning takes place within accepted boundaries; it is adaptive learning that leaves basic values unexamined and unchanged. We all experience this from day to day: learning how to settle household budgets, for example, does not require us to examine or change our values and beliefs. Most learning institutions are primarily engaged in this functional, first order learning where the stress is on 'information'.

By contrast, second order change and learning involves critically reflective learning, when we examine the assumptions that influence first-order learning. This is sometimes called 'learning about learning' or 'thinking about our thinking'. At a deeper level still, when *third-order learning* happens we are able to see things differently. It is creative, and involves a deep awareness of alternative worldviews and ways of doing things. It is, as Einstein suggests, a shift of consciousness, and it is this *transformative* level of learning, both at individual and whole society levels, that radical movement towards sustainability requires.

In any crisis situation, people may stay 'stuck' in first order learning, that is keep 'doing more of the same' which is likely to lead to *breakdown*; or alternatively, achieve *breakthrough*, which is dependent on reflective, intentional learning which gives rise to new perspectives.

The key point is this: the crisis/opportunity of sustainability requires second—and where possible—third order learning responses by cultural and educational systems. There is a *double learning process* at issue here: cultural and educational systems need to engage in deep change *in order* to facilitate deep change—that is, need to *transform in order to be transformative.* To give a brief example, in my experience with WWF-UK's Reaching Out programme of in-service work with teachers on education for sustainability, we found that innovative programmes that encouraged deep personal and professional reflection often changed people's lives.

According to Clark,[7] in the last 2,500 years there have been only two "major periods of *conscious* social change, when societies deliberately 'critiqued' themselves and created new worldviews." From this perspective, 'the learning society' is one that is able to understand and re-direct itself. For us, such social learning will involve taking charge of the evolution of our consciousness at individual and social levels; a deep learning which questions and examines our basic assumptions and values and intentionally speeds the emergence of the *core values* of sustainability such as sufficiency, efficiency, community, locality, health, democracy, equity, justice and diversity. It requires taking charge of our education and learning systems so that they make a positive rather than a negative difference to the human, and indeed, non-human prospect.

An ecological approach

". . . the unhealthiness of our world today is in direct proportion to our inability to see it as a whole."—Senge[8]

According to the ecological viewpoint, the fundamental problem that links the crises we now face is one of inadequate perception. From our earliest years, we are taught to make distinctions to make sense of the world. But our dominant mechanistic worldview takes this to extremes. We reify borders which blind us to the connective and dynamic reality they demarcate: humans/nature; local/global; present/future, cause/effect are prime examples. Our categorization of 'health', 'environmental', 'political', 'economic', 'social' issues and so on, belies the essentially unbroken nature of reality. As the Brundtland Report[9] recognized: "Compartments have begun to dissolve. This applies in particular to the various global 'crises' that have seized public concern. . . . These are not separate crises: an environmental crisis, a development crisis, an energy crisis. They are all one."

But we still think of them as separate—we often fail to see connections and patterns. By contrast, an ecological view of the world emphasizes *relationship.* Such thinking is systemic rather than linear, integrative rather than fragmentary. It is more concerned with process than things, with dynamics than linear cause-effect, with pattern rather than detail. It is both descriptive and purposeful, being concerned with both recognizing and realizing wholeness.

As the issues that surround us are fundamentally systemic, we need to think in an integrative way and act accordingly. The new jargon of 'joined up thinking' indicates a realization of this need. In a growing number of areas, ecological thinking and practice is increasingly evident, including in new economics, holistic science, sustainable agriculture, ecological design, community regeneration and Local Agenda 21 work. The irony is that such movements are way ahead of mainstream education in mapping out paths to a more sustainable future, yet it is education that has been repeatedly upheld as the key to securing sustainable development.

My argument, based on many years of involvement in the field, is that Western education, while founded on a mechanistic paradigm and overlaid by a utilitarian market philosophy, cannot much assist us towards sustainable lifestyles. Furthermore, the reorientation of education towards sustainability is frustrated partly because there is an insufficient vision and elaboration of the basis of such reorientation. I believe that ecological or whole systems thinking offers the potential both to critique current educational theory and practice and to provide a basis by which it may be both transformed and transcended.

A personal perspective

It may be of interest to relate a little of my story. It's early morning, and my young children have woken, and they come into my study. What sort of world will they and their peers inherit? Or their children—can we assure them that the world will be safe and sustaining, even just one generation ahead? Whilst there is no doubt that there is a groundswell of thinking and action towards sustainability, the main indicators, as measured say by the annual Worldwatch *State of the World* reports, or even as reflected in our daily papers, remain deeply worrying. I'm not prepared to just hope it gets better. The future is in all our hands. That's why I'm involved in education for sustainability.

It goes back to my own childhood. There are several episodes which drew me to concern for the environment and for education, including Rachel Carson's seminal *Silent Spring*; an excellent teacher who gave us the seeds of critical awareness; and a television debate between E.F. Schumacher and a status quo economist. Though a child, I had a strong sense that what Schumacher was saying was very important. Skip a few years, and then Paul Ehrlich's *Population, Resources and Environment* made a big impact on me, soon followed by Barbara Ward's *Only One Earth*, written for the 1972 UN Conference on the Human Environment in Stockholm.

Around that time, I came across an American book with an intriguing title, *Teaching for Survival* by Mark Terry. This was one of the first books which not only talked about the need for human ecology in education, but also addressed a second level, which might be termed the ecology of education, meaning seeing the school, its ethos, its curriculum, its community, and so on, as an interrelated whole. It also made the point that all education, whether or not intended as such, is environmental. It is disappointing to have to note that much of what Terry advocated then is still awaiting implementation in the majority of schools. But it was partly through reading that book that I went into education, as I saw it as an essential means by which environmental issues could be addressed.

Over nearly thirty years in environmental education, I've worked in education in many different capacities and in different countries: as a secondary school geography teacher in Sussex; science teacher on a socially-deprived Indian reservation in northern Canada; supply teacher in the prairies; deputy director of the Council for Environmental Education; tutor/course writer for the first MSc in Environment and Development Education in the UK; and freelance consultant to academia, statutory agencies and non-governmental organizations (NGOs) both in the UK and abroad—including recent responsibility for WWF-UK's in-service programme on education for sustainability. During this time I've been fortunate to have worked at all levels, from the classroom to government level, variously as a teacher, writer, researcher and advisor. But I've also been a learner. It is this experience that convinces me that deep change is needed at *all* levels, towards a more whole and integrative view of education and learning.

Over these years, I came to see that the early assumption, shared by most people in environmental education, was a simplistic and deterministic one: that if people learnt about environmental issues, their behaviour would change. Not only does it not work, but too much environmental knowledge (particularly relating to the various global crises) can be disempowering, without a deeper and broader learning process taking place. I still believe that education and learning are absolutely central—and are qualitatively different from all the 'instruments of change' that governments consider in relation to environmental policy like regulation, tax and financial incentives. But realizing the potential of education means recognizing the richness and subtlety both of the learning process and of sustainability, and the dynamic between education and wider society. Education is not a simple 'instrument for change', although good education always involves change in the learner. Engaging education fully in the transition to sustainability requires critiquing much current thinking and practice, but also visioning and designing a credible and practicable alternative—whether you are a policy-maker, lecturer, teacher, community educator or parent.

At the time of writing, 'education for sustainable development' has been recently recognized in the national curriculum for schools in England. I had a lead role in the lobbying and developmental work which led to this welcome change, but I can't help feeling some disquiet. The inclusion of some sustainability ideas such as 'biodiversity', 'carrying capacity' or 'equity' in a curriculum may be an encouraging start, but if 'education for sustainable development' becomes assimilated within a mainstream which otherwise remains unaffected, we shall have achieved little. The challenge is to make it meaningful, and resist the tendency to put it 'in a box'. Going beyond an accommodatory response requires the deeper understanding that ultimately the argument goes way beyond a simple 'add-on' *about* sustainable development. It requires the elaboration of a lived *sustainable education paradigm* which includes, but goes far beyond curriculum, to embrace and suggest a new participative epistemology.

Learning for responsibility requires educational systems, institutions and educators to acquire *response-ability*—the ability to meet the challenge and opportunity that sustainability presents. It necessitates a deeper, more empathetic response to people and to the non-human world. It means putting heart, soul and spirit back into our thinking and practice. Education is not about realizing production but realizing potential, not building competitive league tables but building human and social capacity. Not about merely acknowledging the environment, but understanding that we are deeply enmeshed in its quality and prospects. Realizing sustainable education is a huge but immensely important challenge, but the smallest change can be a step in the right direction, and may affect the whole.

Ultimately, the prospect of a more humanistic and ecological form of education and learning depends on how far all concerned reclaim and engage in the education and learning debate wherever they live and work, and help put it into practice for the common good. As a stimulus to thinking, debate and action, I hope the Briefing helps towards this end.

Stephen Sterling
Dorset, February 2001

Keywords

Education—from the Latin *educare*, meaning to rear or foster, and from *educere* which means to draw out or develop. While this developmental and transformative meaning retains currency, it has largely been overshadowed by transmissive ideas relating to instruction and teaching. 'Education' (as a verb) is commonly used to mean a process, and also (as a noun) shorthand for the 'education system' which involves policies, institutions, curricula, actors etc.

Learning—at simple level, the process through which new knowledge, values and skills are acquired. At deeper level, it involves 'a movement of mind' (Senge, 1990).

Chapter One

Towards Sustainable Education

"The volume of education . . . continues to increase, yet so do pollution, exhaustion of resources, and the dangers of ecological catastrophe. If still more education is to save us, it would have to be education of a different kind: an education that takes us into the depth of things."—E.F. Schumacher[1]

Sustainable education

Western education is presently characterized by a number of paradoxes, which raise some profound questions about its role. Firstly, for nearly thirty years education has been identified in international and national policies as the key to addressing environment and development issues, and latterly to achieving a more sustainable society. Yet most education daily reinforces unsustainable values and practices in society. We are educated by and large to 'compete and consume' rather than to 'care and conserve'. Secondly, education is, as never before, subject to unremitting emphasis on inspection and accountability in the name of 'quality'. Yet dysfunction, stress and the pressure to compete are widely compromising the quality of educational experience and of the lives of educators and learners. Thirdly, governments are concerned about the 'socially excluded', drop-outs from schooling and 'failing' schools and higher education institutions; yet policies which force institutions to compete mean that the advantaged ones get better and richer while the disadvantaged ones become further disadvantaged and receive blame for failing.

The first issue relates to a crisis *of* education, its limited present ability to contribute to a better world.[2] The second and third issues relate to a crisis *in* education, its limited ability to assert humanistic and democratic values in the face of quasi-market and managerialist forces. The two crises are of course related. Policy-makers, meanwhile, understand something by the term 'crisis in education', but it is interpreted in terms of 'failing' students, teachers, schools, colleges or authorities; they don't appear to understand the idea of a 'crisis of education' in the broad sense.

Meanwhile, the environment/development crisis continues, fuelled partly by the human legacy of the last century's educational practices. Clearly, as Schumacher pointed out, "more education" is not the answer to this crisis—or at least, not more of the same. Fundamentally, we need a changed educational paradigm, one that addresses and indicates positive directions beyond these crises, one that "takes us into the depth of things". This is what I term 'sustainable education', a change of educational culture which both develops and embodies the theory and practice of sustainability in a way which is critically aware. This would be a transformative paradigm that values, sustains and realizes human potential *in relation to* the need to attain and sustain social, economic and ecological wellbeing, recognizing that they are deeply interdependent. Ecologically sustainable development depends on sustainable education and learning—which in turn manifests and sustains sustainable development: they are neither separate nor the same. It is an extension of the mutuality nicely summed up in the phrase, "You cannot learn without changing, or change without learning".[3] In the context of sustainability, we are only beginning to understand the full extent of what this means and implies.

Yet in some senses this is nothing new. Many indigenous peoples have been practising forms of sustainable education in their own contexts over thousands of years, maintaining an *intelligence du milieu*,[4] which globally we now lack. But for Western education systems, nothing less than a sustainable education paradigm will do, and time is short to realize this change. As I have stated elsewhere: "Whether the future holds breakdown or breakthrough scenarios . . . people will require flexibility, resilience, creativity, participative skills, competence, material restraint and a sense of responsibility and transpersonal ethics to handle transition and provide mutual support. Indeed, an education oriented towards nurturing these qualities would help determine a positive and hopeful 'breakthrough' future".[5]

Calls for a change in thinking

It has been said on a number of occasions that the world for which education is preparing people no longer exists. As futures educator David Hicks has pointed out, "If all education is for the future then the future needs to be a more explicit concern at all levels of education".[6] But if it's hard to find 'the future' in contemporary education, it's also hard to find evidence of present global trends. Such trends include increases in globalization, telecommunication, technological change, economic and cultural homogenization, interdependence and dependence, complexity, uncertainty, inequity and debt, conflict, consumption, population, movement of people, species loss, destabilization of ecological systems, and climate change. All young people will spend their lives in this century, coping with the profound implications of what has been called this 'world problematique' where each issue in some way is connected with all the others, and cannot be understood or addressed in isolation. A number of studies have shown relatively high levels of *awareness* amongst young (and old) of these issues, but often poor *understanding*.[7] Meanwhile, rapid social, economic and technological changes lead to angst, stress and loss of identity among young and old, who are offered little apart from the sedative mirages of consumption and materialism. Thus, one critical role of education must be to recognize and help people work with these very real concerns and emotions.

Arguably, the root of the 'world problematique' lies in a crisis of perception; of the way we see the world. Accordingly, there are calls for 'a new way of thinking' which would allow us to transcend the limits of thinking that appear to have led to the current global predicament. Examination of descriptions of what the desired 'new way of thinking' might be reveals much use of terms like integrative, holistic, systemic, connective, and ecological.

To re-vision education, we must look beyond the often closed world of education. 'Sustainable education' is only likely to emerge if it can connect with and draw strength from positive cultural change in the wider social context. The roots of a new postmodern educational paradigm are to be found in a number of converging 'growth areas' in wider society which reflect systemic thinking in some way: revisionary postmodernism; the ecological movement and worldview; the sciences of complexity; participatory and ecological democracy; and ecologically sustainable development theory and practice. Keywords from these areas give some idea of what an ecological education paradigm may look like (see **Box 2**).

But perhaps this is going too far, too fast. We need first to look at the current state of education. I begin this by turning attention to the most fundamental of questions, which concerns the purpose of education.

The roles and nature of education

There are three central questions that are key to unlocking the values of any educational system or ideology.

- What is education for?
- What is education?
- Whose education?

To address the first question—What is education for? This most fundamental question is one that often least features in educational debate and policy, or teacher education. The purposes of education are largely taken as given. To ask what education is 'for' raises questions of philosophy and value about the nature of education (the second question), and beyond this, about the nature of being human. Amongst all the current frenetic concern with standards, testing, assessment, and 'quality control', and the struggle of most educators and learners to meet externally determined 'performance targets', these more fundamental questions—which require continuing democratic debate—are pushed to one side.

For many people, however, education is seen as a self-evidently 'good thing', and also as a stock 'answer' to social and other problems—but these perceptions tend to obscure any understanding

Box 2: An ecological education paradigm– some descriptive keywords

Participative, democratic, co-evolutionary, collaborative, reflexive, process-oriented, dialogic, systemic, integrative, connective, adaptive, creative, holistic, synergetic, transformative, purposeful, epistemic.

that differences in educational policy and practice often rest on deeper differences in assumptions and values. To articulate a more ecological educational paradigm, it is important to bring to the surface the values that are implicit, and indeed, dominant, in current educational thinking, and also trace these back, in Schumacher's words, to 'our central convictions'. The question 'what is education for' reveals different ideologies—which have been and continue to be manifested in educational debate, theory and practice.

Let's look at the functions or roles of education. Any educational system tends to be multi-functional, reflecting a mix of aims and objectives. Not least, all education systems attempt to accommodate the tension between maintaining society and reflecting or encouraging change, and often politicians want both. There are at least four main functions, which at different times jostle and often conflict within education policy, theory, and practice. They are:

- To replicate society and culture and promote citizenship—the *socialization* function;
- To train people for employment—the *vocational* function;
- To develop the individual and his/her potential—the *liberal* function; and
- To encourage change towards a fairer society and better world—the *transformative* function.

The simplicity of this list belies the depth of the tensions that underlie debates between these views of education. A first clarifying distinction that can be made is between *intrinsic* values and *instrumental* values. Educational orientations stressing *intrinsic* values view education as an end and a good in itself, as having inherent value and meaning. In this orientation, the use to which the 'educated person' puts their education is a secondary consideration, but it is believed a well-rounded education will only have beneficial social consequences. This was exemplified by the child-centred and 'progressive' movement in education that was at its zenith in Britain in the 1960s.

On the other hand, the *instrumental* stance values education as a means to an end: whether to assist international competitiveness, combat drugs, or promote peace. Hence any phrase conjoining 'education' and 'for' usually implies an element of instrumentalism. There are very many of them as education is often seen as the universal answer to problems. For example, education 'for literacy', 'for employment' or 'for the environment', is seeking a change in the individual or society through education.

This is an important distinction, because an instrumental view of education tends to stress *purpose* and product. It is concerned more with 'what education is for', rather than the nature of education. The intrinsic view however stresses *process*—the quality of experience of teaching and learning, and is concerned with 'what education is' rather than what it might lead to or influence. Both are political, in the sense that the orientation or ends of education (purpose) and methodologies of education (process) rest on deeper value positions. These deeper core values, which help define the underlying worldview of education, are critical and need to be examined if we are to fashion a more ecological and democratic educational paradigm.

Meanwhile, it can be seen that the first two functions of education listed above—*socialization* and *vocationalism*—tend to stress instrumental values, while the *liberal humanist* view of education tends to stress intrinsic values. The *transformative* (or reconstructionist) view, is instrumental in working for change for the better, but often also recognizes intrinsic values and the quality of learning, stressing democratic and participative methodologies. Thus the liberal and transformative views of education are more likely to talk about the 'roles' rather than the 'functions' of education.

Sustainable education is ultimately about reconciling all four views, but particularly builds on last two. It is about integrating and balancing process (what education is) with purpose (what education is for), so that they are mutually informing and enhancing. It builds on the best of existing thought and practice in the liberal humanist tradition, but in many respects goes beyond this. It acknowledges the long held belief in liberal circles that education is about nurturing and realizing inherent potential, *but also* is acutely aware that we need to educate for sustainability, community and peace in a turbulent and rapidly changing world.

Lastly, addressing the third key question ('whose education?'), sustainable education is also democratic education. It seeks to place ownership and determination with educators, learners and communities rather than governments and corporations, and upholds the fundamental value and right of equality of opportunity for all. If we want people to have the capacity and will to contribute to civil society, then they have to feel *ownership* of their learning—it has to be meaningful, engaging and participative, rather than functional, passive and prescriptive.

The modernist agenda in education

At present, an extreme instrumentalism dominates educational policy and practice. In Britain and other Western countries such as the USA, Canada, Australia and New Zealand at least, a very managerialist view of education has come to dominate, modelled on economic change and the perceived 'demands' of a globalized economy and increasingly, globalized culture. This change is not peculiar to the field of education, but 'marketization' has infiltrated virtually all areas of public life including sport, health, the penal system, policing and local government. Government policies have opened the provision, support and monitoring of education to private interests, and corporatization is becoming increasingly enmeshed with state systems, leading to a worrying loss of democratic control and accountability, particularly at local level.

In some European states, such as Denmark, Norway and Sweden, humanistic education traditions are sufficiently strong to have resisted some of these changes. However elsewhere, within some fifteen years, the neo-classical and liberal humanist models of education that informed thinking and practice for the best part of the 20th century have been aggressively challenged by neo-liberal and neo-conservative views (within a socio-economic context deeply affected by globalization, and a technological context increasingly dominated by the nature of internet-based communications). There is a sense that the 19th century model of education that we are most familiar with cannot suffice for the very different conditions and challenges of the 21st century. This sense is a powerful force for change, but the direction that educational thinking and practice is heading in the longer term is yet unclear, and therefore, opportunities may be now arising for more integrative and ecological approaches. In the short term however, technocentric managerialism is king.

In terms of the roles of the educational system, we can understand the current shift as follows:

Vocational function. Now uppermost, with emphasis particularly on skills for the information economy.
Socialization function. Some renewed emphasis as a political response to perceived social problems (viz. 'citizenship' and social inclusion).
Liberal role. Mostly survives in the ethos of individual teachers and some schools and other educational institutions—often in the private sector.
Transformative role. Exists in rhetoric, and probably a minority of educators' beliefs and institutions. But evidence of new energies on the margins of the mainstream.

It is important to understand that the changes that the modernist agenda have wrought are within the bounds of the existing paradigm: what has been termed "change within changelessness",[8] or first order change. **Box 3** indicates distinctions between different orders of change in the educational system.

Box 3: Orders of change

Making adjustments in the existing system (first order change):
Education and school/college improvement—doing more of the same, but 'doing it better'. Emphasis on efficiency.
Education restructuring—re-organizing components and responsibilities in the education system. Emphasis on effectiveness.[9]

Changing the educational paradigm (second/third order change):
Redesign of education system and institutions—re-thinking whole systems on a participative basis. 'Doing better things', and 'seeing things differently'.

So, sustainable education is qualitatively different, and depends on second and, where possible, third order systemic change.

However, the effects of the modernist agenda are not all negative, and some changes could be said to be partly resonant with those that are suggested by a more holistic or systemic model of education; for example, the newer emphases on flexible learning patterns, life-long learning, and the self-management of institutions. Nevertheless, this can be seen as part of the influence of economic rationalization, complementary to flexible working patterns and shifting capital, and designed to ensure a supply of adaptable human resources. The core values of sustainability in these changes are neither implicit or explicit.

For many years there have been alternative educational currents such as environmental education within the educational sea, and many committed educators follow one or several of these movements. I will argue below that environmental education and related movements are not in themselves sufficient to give us a sustainable education paradigm, but they are important currents in the larger and deeper wave of change that is required.

Environmental education—and beyond

Environmental education has occupied my whole professional life. I've written about it, researched it, taught it, and generally fought and lobbied for it for many years. I still believe in it, in all its varied manifestations and whether involving the individual or group, institution or community. It all has a place and validity. But arguably, to date it hasn't been enough to make a critical difference, except to those relatively few people who have experienced excellent programmes.

With the rise of 'education for sustainability' from the early 1990s, I have heard comments that are openly dismissive of field studies and 'traditional' environmental education because they do not relate to sustainability issues or political education. Similarly, a number of recent essays chart the shift of environmental education from concern with 'nature' to 'sustainability'. But this misses the point. Apart from the fact that we can learn about sustainability from studying ecosystems, and benefit from empathetic work in nature, our view of environmental education has evolved and changed over the last thirty years or so. This does not at all invalidate the early and still evolving traditions of nature studies and fieldwork, but puts them in a broader context. At the same time, this critique does point to the fact that much environmental education, however worthy, is not sufficient in itself in our quest for 'sustainable education'.

I am also concerned by those educators who seek to circumscribe environmental education in the face of the challenge and opportunities of sustainability; who strive to maintain existing labels—echoing perhaps the skirmishes of the early 1970s when environmental education was itself the 'new kid on the block'. It's perhaps understandable: the failure of environmental education to permeate all of education, has meant that some environmental educators have adopted a survival strategy of defining their difference rather than their general relevance. The great hope that environmental education would 'create new patterns of behaviour of individuals, groups and society as a whole towards the environment' has been quietly forgotten, even if it was always too ambitious. This was one of the three goals adopted in the Final Report of the First (and as it turned out, only) Intergovernmental Conference on Environmental Education held in Tbilisi in 1977.

Nevertheless, the UNESCO Tbilisi conference was historic. From a standing start, environmental education has grown into a worldwide movement. But while it became broad in scope and extensive in practice, it also became fragmented, away from the remarkably holistic vision that the conference articulated. From a base in rural and local studies in the 1960s, the term and practice of 'environmental education' emerged more strongly in the 70s, embracing also urban issues, and ethical and political dimensions. The global dimension was added in the 1980s, and the parallels with other movements of education for change, particularly development education, were recognized in the 1990s leading to the emergence of 'education for sustainable development'.

The scope of what was considered relevant to environmental education grew ever broader over these decades, while emphases within the field asserted themselves: conservation education, urban studies, nature and environmental studies, critical education, and so on. But both tendencies—expansion and fragmentation—made it difficult to demonstrate and communicate intellectual coherence to policymakers.

At the same time, a series of parallel movements concerned with education for relevance and social change emerged, including development education, peace education, world studies, anti-racist education, human rights education, human-scale education and holistic education. Attempts were made to subsume these forms of 'education for change' under the aegis and philosophy of *global education* in the 1980s, and this term retains some currency as an integrative vehicle. Increasing recognition by practitioners that their concerns were mutual and relative, and greater awareness of the challenge of sustainability helped the emergence of the concept of *education for sustainability* in recent years as an alternative catch-all term. This term is preferred by some radical educators to 'education for sustainable development' but recent evidence shows it is still susceptible to accommodation by the mainstream, which is why I suggest the larger term 'sustainable education', which is much less so.

Some of the implications of sustainable education are examined in the next chapter, but for now, we are left with the conclusion that environmental education and related fields are 'necessary but not sufficient'. If we take an evolutionary view, sustainable education does *not* mean we should abandon the name or the established practices of environmental education and sister fields. Indeed, the multiple perspectives afforded by these fields are valuable. It does mean however, that we extend our thinking and practice, that we value and recognize the links between the various forms of education for change, and that we recognize that these forms of education are themselves in flux. In this way, environmental education and its sister movements do not become 'closed systems', but seeds for broader educational and societal change.

Many practitioners in environmental education are disappointed at its relative lack of impact over the last decades. But we have to recognize that its values, theory and practice are affected, influenced and constrained by the systems within which it is embedded, these being the broader educational system, and in turn, this within the larger social system. Therefore to understand the possibility of, and barriers to, a sustainable education paradigm, we must first attempt to understand the 'ecology of education' in terms of system levels and relationships.

The 'ecology' of educational systems

One of the important concepts and tools in systems thinking (see Chapter 3), is that of *nesting systems*. According to this theory, which derives most notably from Koestler's[10] idea of 'holons', reality can usefully be modelled as a hierarchy of nesting systems, where the bigger context (suprasystem) shapes, limits and gives meaning to the smaller part (subsystem), rather like the analogy of the Russian doll.

Using this insight, movements for educational change such as environmental education can be seen as subsystems of the larger or mainstream formal educational system. In turn, this educational system can be seen as a subsystem of the larger socio-economic and cultural systems, which also directly 'educate' people. Socio-economic systems must be regarded as subsystems of the encompassing biophysical system.[11] (The fact that the economic system is often seen as independent of, or encompassing, the biophysical system is partly the root cause of our current crisis, of course.)

This model shows that the central question that has been occupying environmental educators for decades—'How can environmental education change people's regard for and behaviour towards the environment?'—is misplaced, or at least over-optimistic. The linear idea that more environmental education would change people, and thereby would change society, ignores at least three realities:

- education for change is often outweighed by the larger educational system which enacts vocational or socializing roles and purposes, and can 'cancel out' radical educational endeavour;
- the larger-still social system affects and shapes the educational system more than the other way round, although they are in a dialectical relationship; and
- in an age of mass communication, the socio-cultural milieu arguably affects people and influences values more than formal education programmes do.

These dynamics do not mean that education for change movements cannot be effective, just that they are always limited by factors beyond their influence.

The systems perspective encourages a change of question, to 'How can education and society change together in a *mutually affirming* way, towards more *sustainable patterns* for both?' This is a change of focus that allows a more creative and subtle response. In systems terms, it is seeking a 'positive feedback loop' whereby change towards sustainability in wider society supports sustainable education, which in turn supports change in wider society, and so on. It takes us from a model of education as one of social reproduction and maintenance, towards a vision of continuous co-evolution where both education and society are engaged in a relationship of mutual transformation—one which can explore, develop and manifest sustainability values. In immediate terms, it means working to make the educational institution as far as possible a microcosm of the emerging sustainable society, rather than of the unsustainable society.

Is this realistic? The concept of sustainable education appears to be calling for deep change at a time when educators and learners are already overwhelmed with too much change. But sustainable education is of a different order: affirming rather than dislocating, hopeful rather than soulless, where the smallest gain can be of deep significance. But to make this effective we need to understand the nature of the changes that have been imposed on formal education in recent years, and contrast them with the sort of change that a participatory ecological education paradigm implies. This is the theme of the next chapter.

Diagram: Nesting systems

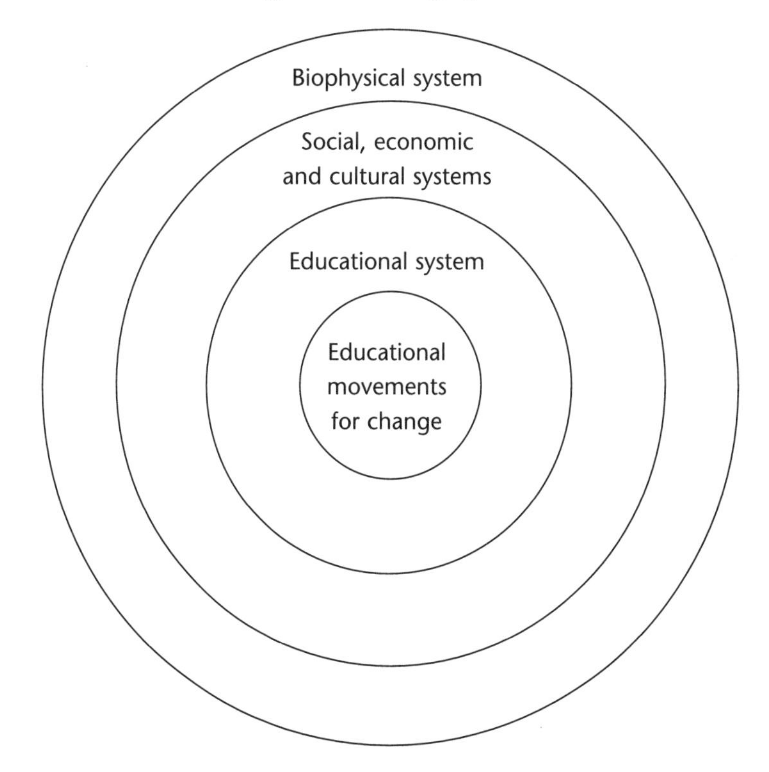

Chapter Two

Education and Learning in Change

"What is needed is a radical change in perspective within educational institutions to deal with the magnitude of the problems that we are currently facing at a planetary level."—O'Sullivan[1]

Understanding education and change

Education has been in the political limelight for a long time. In the UK, this can be traced back to Prime Minister James Callaghan's launch of 'The Great Debate' about education in the late 1970s. Formal education has been endlessly reviewed, criticized, reformed, judged, assessed and reshaped in the period since. Current policy, as we have seen, stresses vocationalism over everything else; education is seen primarily as the key to economic competitiveness, although recently, and with mounting concern over social issues, the concepts of citizenship and social inclusion have been introduced as a socializing influence. This balancing act, which tries to promote both education for economic change and education for social stability, is a current tendency in Western European education systems.[2]

But much of the educational debate and policy change is confused and confusing. To clarify the relationship between change and education, it is useful to distinguish between 'education for change' and 'education in change'. The relation between these two meanings is important, and is a key theme of this chapter.

- Education *for* change concerns the role of education in bringing about change in the person or society. It is about change sought or achieved *through* educational practice.
- Education *in* change refers to the policy changes made to educational rationale, theory and practice that affect and may facilitate (or hinder) education for change.

Essentially, all education can be seen as expressing either a broadly 'transmissive' or 'transformative' methodology (methodology goes beyond 'teaching method' to also reflect philosophy and purpose of education).

With a transmissive methodology, 'education for change' is *instructive*—i.e. associated with the transfer of information—and 'education in change' is *imposed.*

With a transformative methodology, 'education for change' is *constructive*—i.e. engages the learner in constructing and owning meaning—and 'education in change' is *participative.*

'Imposed' and 'participative' describe the style of policy change and management, and apply to any level of the education system. This model gives us a set of descriptors which can help us understand the bases and kinds of educational change (elaborated in **Table 1** and discussion below). Notice that 'education for change' and 'change in education' are necessary to each other: the former cannot be achieved without sympathetic change in the latter, and vice-versa. A further point to note is that in practice, both categories of change are usually matched, reflecting either transmissive or transformative styles. My argument here is that sustainable education is essentially transformative, constructive, and participatory.

The problem of education for change

Let's start by looking at education for change in relation to sustainability. Numerous international statements and mandates have pointed to the key role of education as a change agent: from creating "new patterns of behaviour of individuals, groups and society as a whole towards the environment",[3] to being "critical for promoting sustainable development and improving the capacity of the people to address environment and development issues".[4] Agenda 21 talks of the need to 'reorient' education towards sustainable development. In the European Union, there is the 1988 Resolution on Environmental Education, and the Environment Programmes which put store by education and training in achieving sustainable development.

Yet by the last UNESCO international environmental education conference in 1997 in Thessaloniki, UNESCO was reflecting on why the world's education communities had not responded to these clarion calls. As UNESCO's then Director-General commented,

"Who would deny that too little has been achieved?",[5] while a recent report to the Commission on Sustainable Development indicates that the major work is still to be done.[6] Similarly, the goal of UNESCO's 1990 Jomtien conference on Education for All (EFA), which promised basic education for all by 2000, had not been met by the time of the World Education Forum held in Dakar in April 2000.[7]

There is a crisis here, and a major reason seems to be that member states are not interested in 'education for change' of this sort, but rather of a different kind, relating to the global economy. As the current UNESCO Director-General has suggested, "Education for all has often been seen as a burden by governments trying to adjust to the demands of global competition."[8] But a further reason why educational systems across the globe have hardly responded to the challenge of reorientation may be that there has been insufficient explication of the changes *in* education that would be necessary for educational practice to fulfil the international rhetoric. What has been missing is clarity about the vision of education for sustainability that is needed, and also a strategic sense of how progress towards such a vision could be made, bearing in mind the power of the dominant social and educational paradigms. The conjunction and mutual informing of the *visionary* and the *practicable* is essential in successful change, at any level.

'Education for change' is also the emphasis of most NGOs involved in social, economic and environmental issues. Education is seen as a means to raise environmental awareness, improve literacy, primary health and quality of life. NGOs, less bound by centralized education policies, and often operating within community education settings, have often found that successful education *for* change both in the Southern countries and in the developed North, depends on participative change *in* education—that the two are co-determining. It was not always the case: using participative methods, respecting existing knowledge, recognizing local conditions and culture (as practised in such context-related approaches as Participatory Rural Appraisal), had to be learnt from the failure of more transmissive and imposed methods.[9] Such constructive methodologies are more difficult, time-consuming, and unpredictable, but change is owned by the participants, and therefore more likely to be sustainable.

Governments are also interested in education for change. Drugs, smoking, sex, racism and AIDS are all examples of areas where governments often want to make a change in behaviour through 'public education'. The problem is that they very largely choose transmissive and instructive strategies, often based on information and campaigns. The basic assumption is one of deficit—that something is missing or wrong with people. The emphasis then is in 'getting the message across', rather than engaging with people and their expressed needs. Despite numerous studies showing the often superficial and impermanent impact of such programmes, they remain popular with authorities. Further, the instructive approach of 'information transfer', particularly since the domination of neo-liberal thinking, is the often prime *modus operandi* in formal education as well as in non-formal arenas.

It is not that instructive approaches are always inappropriate. Training in certain skills is entirely sufficient in some aspects of vocational training, for example. But even here good training becomes an applied art rather than a universal science, and will often extend into constructivist approaches. Situations that require 'higher order' or transformative learning always imply constructivist approaches, and this is very largely the case with sustainability issues because of their complexity and often deeply challenging nature.

Table 1 summarizes the differences in approach and values between transmissive and transformative education.

If education for change is problematic, then policy-led 'change in education' should be of even greater concern because this ultimately shapes practice, and because of the seismic shifts that have taken place in this area in recent years. In other words, the chances of transformative education practice also depend on understanding what is happening in the area of policy, and being able to envision a participative alternative.

The problem of education in change

"Most secondary schools are unhappy places. There is so much wrong now that a total re-think is required."—Senior secondary school teacher, Southampton, UK

Formal education has been under such unprecedented pressure of change recently that the stresses are all too apparent. Many educators feel that yet more change is itself unsustainable. Too many good teachers, lecturers and heads in the formal sector are taking early

Table 1: The differences between transmissive and transformative education

	TRANSMISSIVE	TRANSFORMATIVE
EDUCATION FOR CHANGE (Practice)	**Instructive**	**Constructive**
	Instrumental	Instrumental/intrinsic
	Training	Education
	Teaching	Learning (iterative)
	Communication (of 'message')	Construction of meaning
	Interested in behavioural change	Interested in mutual transformation
	Information—'one size fits all'	Local and/or appropriate knowledge important
	Control kept at centre	Local ownership
	First order change	First *and* second order change
	Product oriented	Process oriented
	'Problem-solving'—time-bound	'Problem-reframing' and iterative change over time
	Rigid	Responsive and dynamic
	Factual knowledge and skills	Conceptual understanding and capacity building
EDUCATION IN CHANGE (Policy)	**Imposed**	**Participative**
	Top-down	Bottom-up (often)
	Directed hierarchy	Democratic networks
	Expert-led	Everyone may be an expert
	Pre-determined outcomes	Open-ended enquiry
	Externally inspected & evaluated	Internally evaluated through iterative process, *plus* external support
	Time-bound goals	On-going process
	Language of deficit and managerialism	Language of appreciation and cooperation

retirement or changing jobs. In England in 2000, a record number of heads—some 2700—left the profession or took early retirement. In the UK, in recent years some teachers and pupils have even taken their own lives due to stress levels. There is a dire shortfall of recruits to the profession. For many committed teachers the story is the same: the changes have often been profoundly demoralizing, and anti-educational in the liberal sense. Many aspects of the current drives to change educational culture are in their effect pathological, although of course they are not intended as such. We need to understand what's happening if we are to move towards a more humanistic and ecological way of thinking about, and practising education.

In sum, the New Right, or neo-conservative and neo-liberal forces, have captured and changed the educational agenda. This has taken place in a number of Western education systems. This powerful momentum appears to have begun through the influence of international agencies such as the OECD and World Bank, which from the 1980s produced reports outlining the 'need' to restructure education in parallel with economic 'structural adjustment'. Endorsed by both conservative and liberal governments, this new culture blamed schools and teachers for failure (in economic terms), but at the same time saw them as a key to maintaining economic growth and competitiveness in the new enterprise culture.

Changes have affected everything from the purposes of education, to curriculum, funding, management, the role of schools and teachers, and even the overall ethos of education. The emphasis has shifted from *educational* values to do with process, and developing potential and autonomy, and *social* values relating to equality of opportunity, community and social cohesion, towards *economic* values such as efficiency, quality control and production, which education is required to serve in an unprecedented way. This has affected how teachers, learners and schools see themselves, thus defining a new marketized set of relationships. Thus "teachers are constructed as 'providers', principals as 'managers', parents as 'employers', and students as 'consumers'".[10]

The main changes have been: centralization of control over the curriculum; weakening of local authority support and control; weakening of networking and collaboration between teachers and between educational institutions; encouragement of traditional teaching methods; heavy systems of inspection; competition between institutions to gain pupils/students and resources; precise control over teacher education; and turning heads and principals into managers. The analogy with the factory is telling: young people and qualifications are produced; there are precise goals and targets; the curriculum provides directives for each stage of production; and teachers are technicians and are therefore substitutable. And workers are not required to think too much.

In addition, both in parts of Europe and particularly in North America, the market has come to schools (and colleges) in the form of private sponsorship which has become increasingly essential rather than supplementary to school budgets and led to new inequities between the schools that attract corporate backing and those that do not. And it has also led to increasing corporate influence on curriculum and administration, and a lexicon that includes 'the education industry'. This trend is increasing: for example, the European Union recently decided that every publicly-run school in Europe must be twinned with a corporation by the end of the decade.[11]

The managerial approach in education reflects mechanistic beliefs in determinism and predictability—which leads in turn to a belief in the possibility and merits of control. Arguably, the agenda is not so much about standards, as *standardization*. This is manifested, for example, in the emphasis on precise 'learning outcomes' and 'performance indicators'—if we know what the inputs and process are, we can say with certainty what the outputs should be, and can judge success and failure on the basis of how far predetermined outputs are achieved. This way of thinking about education has become part of the shared psyche, ousting previous and perhaps more valid ways of

Bob Davis, a retired Toronto high school history teacher, worries that with the emphasis on education on how to do various things, the larger context of what is going on in the world at large is being lost. He finds that many kids come into the classroom generally confused and "mixed up about their basic beliefs". Davis bemoans how language skills and mathematics have replaced basic subjects like literature and history in Canadian schools. "There is more focus on training for work and less on how to live in this society and how to live in the political system."[12]

perceiving education and learning, and narrowing perceptions of what constitutes worthwhile knowledge and enquiry.

While there have arguably been some positive aspects (reviewed below), the effect of the managerialist revolution has too often corroded what many teachers would uphold as 'good education' and an atmosphere of co-operation and creativity where learning can flourish. In an age where everything has to have quantitative indicators if it is to count for anything, everybody knows somebody in teaching or learning who can recognize some of the following tendencies in most formal education settings:

- a narrowing of what counts as achievement to that which can be measured;
- high stress amongst teachers/lecturers and a feeling of being out of control and undervalued;
- deprofessionalization of teachers/lecturers, who become technicians rather than reflective practitioners;
- a breakdown of a sense of community within the school/college and of professional collegiality;
- return to limited learning styles and didactic teaching;
- decline in teacher-led innovation;
- marginalization of the arts and humanities subjects;
- equation of quality and achievement only with measurable results;
- a creeping up of marks over time as it is in everybody's interests to demonstrate 'high standards';
- less sharing of experience between schools in a locality;
- increasing gulf between 'have' and 'have not' schools;
- less time and ability to respond to differentiated needs amongst pupils, and to pupils with social or emotional difficulties;
- breakdown of community links as children travel miles (usually by car) through 'parent choice';

- valuing what can be measured, rather than measuring what is valued;
- weakening of local authority support; and
- difficulty in recruiting to the profession.

Moreover, the concept of 'quality' has been hijacked. This is increasingly questioned by many working in higher education (HE). For example, Bassnett, a university pro Vice-Chancellor, argues strongly that the obsession with 'quality' is leading to the opposite, a loss of quality if a wider set of quality criteria is taken into account. The amount of energy and time that goes into bureaucracy and administration to "prove you have demonstrable quality learning outputs" she argues, has led to loss of good teaching and research, loss of tutor-student contact and the growth of "a whole new cadre of university bureaucrats living comfortably off the quality industry".[13]

Laurillard, of the Open University in the UK, points out that none of the assessment processes that HE now has to follow enable the HE sector to *itself learn:* "they merely describe, and at a level of description that does nothing to help us understand whether we are actually serving our students better".[14] Yet the insistence on measuring, on accountability is everywhere in formal education systems. Tate (a former Chief Executive of the Qualifications and Curriculum Authority in England), has stated, "As a society we are preoccupied with assessment. Never before have so many been tested, for so long, and under such scrutiny."[15]

But why? The craziness of all this is that it is in the name—apparently—of 'driving up standards'. Even if this is the case (and standards are measured narrowly) it is at huge cost. In England, irrespective of the £100 million that the inspection office OFSTED costs to run each year, 'driving up standards' has too often meant driving up stress and disillusionment, driving down morale and the breadth of what counts as worthwhile learning, driving out good teachers/lecturers from the profession and the students that don't make it, and indeed, driving more children to schools miles from home.

What is so wrong about this obsession is that it flies in the face of what good teachers have always known: that young people and in fact everybody learn and work best with recognition, trust, support and encouragement, time and space—rather than fear and high stress. From 'support' at one end of the spectrum to 'judgement' at

"We are used to hearing our schools assailed by critics who want to know why 'Johnny can't read, Johnny can't write' and who call for a return to 'the basics'. . . But why do we stop worrying there?. . . Why not worry that Johnny can't dance, can't paint, can't breathe, can't meditate, can't relax, can't cope with anxiety, aggression, envy, can't express trust and tenderness?. . . that Johnny does not know who he is . . . ? Let us admit that the basic skills have nothing to do with Johnny's health, happiness, sanity, or survival, but with his employability. Whose interest, then, is Johnny's education serving?"[16]

the other, the pendulum has nearly swung off the scale, and at all levels of the system. The constant 'naming and shaming' culture of seeking 'failure' has aggravated failure problems rather than cured them. At the same time, the successful students and institutions are those who are able to play the system—but this is nothing to do with developing reflective and creative thinkers, or responsible citizens.

Seen from a systems or ecological point of view, the negative effects of the managerial paradigm are not surprising, because policymakers are thinking too narrowly, deterministically and mechanistically. What they may think of as unfortunate side effects are in fact emergent properties of a poorly designed and rigid system. Education and learning need to be grounded in the qualities of relationship rather than product, but this truth is lost in the present climate. One of the key ideas in the emerging study of complexity is that complex systems cannot be 'controlled', yet control remains a basic tenet of most policymakers.

From a sustainability point of view, the problem with the current change culture driven by Western governments is that it is first order change, and for the wrong reasons. Their approach to 'education *in* change' is to adapt educational policy to what they predict the globalized information economy will require. Their approach to 'education *for* change' is to educate people to adapt to change (first order learning) rather than develop their capacity to shape it. Critical debate that lies outside or questions this approach is dismissed or ignored.

Clark argues that education "is and ever must be a political endeavour. It either moulds the young to fit in with traditional beliefs, or it critiques those beliefs and helps create new ones".[17] In these terms, the New Right agenda is 'mould to fit'—albeit to fit changing social and economic circumstances. Sustainability, however, requires what Clark calls 'create/critique' education to help human systems work *with and within* Earth's ecological systems: a very different orientation, and one that stresses community, creativity and capacity building rather than control, fit, and dependence.

The modern-postmodern transition

In questioning the managerialist changes that have been imposed on education, I am not advocating a return to some mythical golden age of education, nor suggesting that all the changes associated with the social and economic transition from modernity to postmodernity are deleterious either in educational or sustainability terms. A number of significant changes in the educational landscape have occurred and continue to occur, and some aspects of these have positive potential seen from a sustainability perspective. These changes include more emphasis on:

- learning than teaching;
- life skills and 'life-long learning';
- hybrid and multidisciplinary subjects;
- information technology as a learning and 'delivery' tool;
- distance and open learning;
- 'the learning organization'; and
- recognition of the transitory nature of knowledge.

Meantime there have been a number of recent reports that seek to speed up this process of change.[18, 19] The discourse is about the need for more flexibility in learning and schooling, about getting ready for the information revolution, about life-long learning, about learning to learn, and so on. As in many ways formal education is still largely based on the 19th century factory model, some of these changes might appear welcome. However, this welcome must be qualified if these calls seek to replace rigidity and hierarchy with more effective learning structures simply in order to serve the global economy more efficiently. Ideas of 'the knowledge society', 'the information economy' and 'the information society' hide questions about what sorts of knowledge, controlled by who, for whom, and for what purposes. As well as the problem of sheer information overload in the digital age, there is misinformation, disinformation, and irrelevance. "What is worth knowing?" should be the constant question.

The main thrust of reform is about skills and competencies—but informed by what values? Education is in transition, but towards what, and to what purpose? Again, we need to ask what we want education to be for. The current changes are largely about moving education suited to the modern industrial age to one appropriate to the postmodern information age. But without asking deeper questions concerning *ethos*, and posing the sustainability context, they may exacerbate rather than contribute to the sustainability issue. Without an ecological understanding, we are in real danger of creating post-modern learning institutions, whose graduates are able to exploit others and the environment more efficiently and effectively than their predecessors.

In sum, some of these changes may be not be inimical to a sustainability paradigm, but genuine movement towards sustainable education—rather than just 'more efficient' education—requires understanding of the emergent holistic paradigm in Western culture. I call this the 'postmodern ecological paradigm', and this is briefly examined in the next chapter.

Control, management and business

The glaring irony of the managerialist approach to education is that the business world it is mimicking has moved on—at least in part. There is a considerable time lag here: education-*follows*-business-*follows*-science. But now, while progressive business is beginning to echo a changing scientific paradigm based on complexity, education managers and policy makers still rehearse yesterday's management paradigm believing it to be 'modernizing'.

But this management paradigm is based on the older modernist scientific paradigm, which espoused cause-effect determinism, predictability, control, and objectivism. By contrast, the new sciences of complexity, concerned with emergence and chaos in non-linear systems, indicate that we live in a fundamentally participative world that is both unpredictable and inherently creative.

The application of complexity theory is leading to a new language in management which is questioning the validity of long-term planning, of heavy top-down management, and rather sees management as a subtle art of working with dynamics and relationships. The emerging metaphor is the business or organization as a *living system,*[20] rather than as a machine, and this has profound consequences for the way it is run. This includes, but goes beyond, 'greening' an institution's energy use or purchasing policy—and yet even this is a far goal for most education institutions.

In essence, the shift is from 'systematic control' towards 'systemic learning'. A newsletter of a leading management training institute commented on the implications of complexity theory: "Leaders and managers should aim to develop conditions in the organization which allow self-organizing behaviour to flourish. This means creating adaptive organizations with flexible structures, skills, processes and information flows. Instead of hierarchically imposing change, managers need to unleash the potential for change."[21]

In other words, too much control—exactly the management paradigm being practised in most education institutions—inhibits innovation, and results in rigidity and inability to respond creatively to a changing environment. By contrast, management for complexity values and emphasizes genuine participation and collaboration, flexibility, trust, inclusivity, diversity, autonomy, and the role of local and personal knowledge as inherent to the learning process. And it tolerates uncertainty and instability, as necessary to self-organization, group learning, and innovation.

Leading businesses are beginning to put into practice a changed educational and learning paradigm where the organizations themselves are creative *learning communities,* and where people are valued as the prime resource. Further, they seek employees who can fully participate in a learning culture, who are adaptive, creative, and flexible—rather than the 'cogs' produced by mechanistic education systems and by narrow vocationalism and specialization. These businesses value *diversity*, and see it as essential to their own sustainability.

Such businesses are developing new paradigm ideas based on the primacy of *relationship* in their management and learning practice, and are looking for skills in coping with complexity. Yet they recruit from formal education institutions that largely conform to

Table 2: Essential management differences between mechanistic and ecological models of education

MECHANISTIC MANAGEMENT	ECOLOGICAL MANAGEMENT
STYLE OF MANAGEMENT	
Goal oriented	Direction oriented
Product oriented	Process oriented
Controlling change	Facilitating change
Focus on single variables & parts	Focus on sets of relations and the whole
Aware of causal relationships	Aware of emergence
Power-based hierarchy	Leadership and self-management at all levels
Command and control	Democratic and participative
Vertical structures	Flatter and integrated structures
Intervention from 'outside' system	Working with and from inside system
Interested in prediction	Interested in possibility
Problem solving	Problem reframing and situation improvement
Adaptive learning	Adaptive, critical and creative learning
External evaluation	Self-evaluation with support
Quantitative indicators	Qualitative and quantitative indicators
Planning	Design
Closed	Open
EFFECTS ON SYSTEM (TENDENCIES)	
Standardization	Diversity and innovation
Homogenization	Heterogeneity but coherence
Dependency	Autonomy-in-relation at all levels
Externally directed	Self-organization
Dysfunctional emergent properties	Healthy emergent properties
Poor ability to respond to change	Flexibility and responsiveness
Unsustainability	Greater sustainability

mechanistic thinking and promote uniformity. Sustainable education, however, would nurture the human qualities that progressive businesses and organizations interested in social, economic and ecological sustainability, are now seeking.

The most urgent challenge is for educational institutions to make a conscious shift from their guiding metaphor of 'factory', and move to the metaphor of 'living system'; to move from just seeing themselves as 'teaching organizations', to becoming *learning organizations.* Some of the essential differences between these mechanistic and ecological models, in terms of management, are suggested in **Table 2**.

There are two important dimensions to be added here. Firstly, the question of scale. Arguably, the ecological model of management is only fully possible where the scale of an organization or institution, or of its subnetworks, is conducive to it. The sheer size of many educational institutions, and of the groupings within them, mitigate against human-scale relationships and healthy emergence.

Secondly, the question of context. There is a real danger that educational institutions (and indeed business organizations) embrace *some* of the ideas of complexity and ecological management, yet still remain driven by mechanistic values, and with no sense of the wider context of the need to achieve sustainable lifestyles. Hence, we now look further at the bases of ecological thinking and its implications for education and learning.

Chapter Three

Towards an Ecological Paradigm for Education

"If everything is intimately connected, then the quality and integrity of all kinds of relationships are of paramount importance."—Elgin[1]

The power of paradigms

Arguably, the fundamental tension in our current age is between a *mechanistic* and an *organicist* way of viewing the world. From the ecological perspective, the mechanistic root metaphor is becoming increasingly untenable. But while there is evidence of an emerging cultural paradigm which may broadly be described as ecological and postmodern, there is no certainty that it will prevail, which is why it needs to be better recognized and more widely understood. According to Capra,[2] it reflects a "new perception of reality" which has "profound implications not only for science and philosophy, but also for business, politics, health care, education, and everyday life." Increasing numbers of writers are pointing to the emergence of this ecological worldview, variously called 'participative', 'coevolutionary', 'living systems', or 'New Environmental Paradigm'.

What we are seeing is the emergence of a fundamentally different story about how the world works. Like any paradigm, it can be understood through looking at three key dimensions: a normative aspect (*ethos*) which affirms beliefs and courses of action, a descriptive aspect (*eidos*) which is how we conceive the world, and a practice aspect (*praxis*) which represents manifestation and action. This three-part model applies both to the cultural paradigm, and to the potential educational sub-paradigm which is suggested and outlined below.

The ecological worldview

Ecological thinking entails a shift of emphasis from relationships based on separation, control and manipulation towards those based on participation, empowerment and self-organization. Thus, while we might look back at the 20th century as the age of fragmentation, witnessed in everything from the splitting of the atom to the atomization of thought and knowledge themselves, this new century might yet be the age of relation. Indeed, some writers believe we will see a shift from the modern age of the 20th century based on the metaphor of the machine or mechanical system, to a postmodern age in the 21st based on the metaphor of the organism or living system, suggesting a changed view of reality.[3, 4]

This participative worldview is based on the idea and intuition that we are deeply enmeshed in a reality which is both real *and* created, and that these are inextricably linked: that how we see the world shapes the world, and this in turn shapes us. This is why it is sometimes called 'co-evolutionary'. Because of this unavoidable dialectic, the quality of our individual and collective perception is critical. We need to discover more adequate ways of thinking about ourselves and our relationship with the world through a new, partly rediscovered epistemology. The 'ecological paradigm' represents the expression of this movement and search.

Evidence of this emergent paradigm can be seen in aspects of ecological thinking; in particular, ecophilosophy and deep ecology, social ecology, ecofeminism, transpersonal and eco-psychology, creation spirituality, and holistic science, as well as more practical expressions in areas such as ecological economics, sustainable agriculture, holistic health, and ecological design and architecture. As mentioned in Chapter 2, it is also increasingly evident in businesses' interest in complexity theory in relation to organizational change. However, the education world is largely unaware of this *zeitgeist,* and of its implications for a new education and learning paradigm.

Educational subparadigms

In educational literature, various paradigms are aired about the nature of education, its purpose, and methodology. But the important question is how far all these views are *sub*paradigms, more or less conditioned by the ghosts of mechanism, positivism, and dualism, and the assumptions of modernism within the broader cultural paradigm. 'Deconstructive postmodernism' is supposed to help us here, but it leaves us drifting in a sea of relativism. Thus a number of voices in the debate about paradigms are searching for a postmodern, ecological alternative that is more adequate and creative—and which gives us a basis for action.[5] This is where we need to ground sustainable education.

The key challenge is to create and articulate an educational *ethos, eidos* and *praxis,* based upon the emerging ecological paradigm in wider society. This can be approached through a simple 'whole systems thinking' model, which offers a profound way to help us reorient our worldview, and also our educational thinking and practice.

Vision, image and design

Radical change in education may be seen as depending on developing three related bases,[6] which echo the three dimensions of paradigm outlined above:

- A *vision,* that is, a philosophy and direction;
- An *image* of the desired state in terms of core values and ideas as a basis for discussion; and
- A *design* that allows realization of the image.

Arguably, none of these is sufficient in themselves, but together represent the potential to effect significant change. There is nothing mysterious or in fact ecological about these bases, and the dominance of education by the vision and design of the New Right in recent years perhaps attests to this. But if these three dimensions can be elaborated from an ecological perspective, it gives us a basis from which the dominant and conventional education paradigms can be *evaluated,* and *re-visioned.* This of course must be a debate involving all interested in the future of education.

The constructive vision is one of sustainable education and a sustainable society in mutual and dynamic relationship. This is easily stated, but much harder to elaborate. The philosophical basis of this *vision* is briefly suggested below in terms of whole systems thinking and what I have called 'the connective pattern'. This basis is then used to *image* a suggested outline of an ecological education paradigm which can applied at any level—from learning institution to national system.

Whole systems thinking

Whole systems thinking is a name given to the quality of thinking and being that appears necessary in order to go beyond the dominant forms of thinking which are analytic, linear, and reductionist. Through drawing on systems and humanistic ideas, it offers a way of making holistic thinking understandable, accessible and practicable. Because of this, it has great (but largely unrealized) educational potential. It identifies three interrelated dimensions of paradigm—perceptual, conceptual and practical—which describe human experience and knowing at any level—personal, group or whole societies.

In the dominant paradigm, there tends to be dis-integration within and between these dimensions. For example, in the Western tradition, intellectual knowledge (*conceptual dimension, or 'eidos'*) has primacy, to the extent that other forms of knowledge such as 'intuitive knowing' (*perceptual dimension, or 'ethos'*) or 'practical knowing' (*practice dimension, or 'praxis'*) are often regarded as having less value.

By contrast, the whole systems model provides a basis for understanding the emerging ecological paradigm, wherein each dimension of knowing is extended, deepened and integrated. The necessary shifts from mechanistic thinking towards ecological or whole systems thinking are represented and summarized in **Box 4**.

The guiding principle here is *wholeness*, in relation to purpose, to description, and to practice. Applied to educational settings, this three-part model is congruent with 'values, knowledge and skills', but it will be appreciated that what is implied here relates to a deeper level of transformative change than is usually meant by these words. From a simple concern with values, education needs to heighten awareness of worldviews. From concern with promoting knowledge (and often factual knowledge at that), the shift needs to be towards developing critical and systemic understanding and pattern recognition. From concern with functional skills we need to develop broader and higher order capabilities. The key assumption in this approach remains that we need to 'see' differently if we are to know and act differently, and that we need learning experiences to facilitate this change of perspective. This, as argued earlier, is second and third order learning—of which policy makers and practitioners need some experience if the education system is to also respond. Before exploring this further, I

Box 4: The necessary shifts from mechanistic to ecological thinking

Perceptual dimension—the need to widen and deepen our boundaries of concern, and recognize broader contexts in time and space (Extension).
In an age of individualism and materialism, we are not encouraged to include 'the other' in our thinking and transactions, whether this be neighbour, community or minorities, let alone distant environments, peoples, and non-human species, or 'the needs of future generations'. By contrast, we need an inclusive rather than exclusive view, which recognizes that human and natural systems (and people) are in some way co-dependent and co-determining. As well as the outer dimension of extension, this disposition also requires an inner 'deepening' process, which values all aspects of personhood, particularly intuition, and becomes aware of our individual and shared needs and worldviews. The key quality here is 'empathy'.

Conceptual dimension—the disposition and ability to recognize and understand links and patterns of influence between often seemingly disparate factors in all areas of life, to recognize systemic consequences of actions, and to value different insights and ways of knowing (Connection).
The intellectual ability to know the world in a more ecological or relational way is more likely to give rise to respect in understanding and wisdom in action. The key quality here is 'understanding'.

Practice dimension—a purposeful disposition and capability to seek healthy relationships recognizing that the whole is often greater than the sum of the parts; to seek positive synergies and anticipate the systemic consequence of actions (Integration).
Emergent properties in any system may always surprise us, but they are more likely to do so in positive rather than negative ways if we think, design, and act integratively and inclusively. The further aspect of integration is the awareness of the need to live within the Earth's physical limits. The key quality here is 'wisdom'.

want to suggest that the view of reality that the ecological paradigm is giving rise to, indicates a 'connective pattern' that links learning, education and sustainability.

The connective pattern

"Break the pattern which connects the items of learning and you necessarily destroy all quality. . . . Why do schools teach almost nothing of the pattern that connects?"—Bateson[7]

The insights of whole systems thinking allow us to suggest the essence of the 'pattern that connects' education and sustainability. At the heart of it is *wholeness* and *health* (both words having the same semantic root). These are hard words to define because they are qualitative, but they invoke the ideas of integrity, of the unfolding and maintenance of creative potential in a dynamic state, of an aesthetic and of quality.

Complexity theory and our knowledge of living systems is confirming a widely-shared intuition: that healthy, sustainable systems are those which are self-organizing, self-healing, and self-renewing, and that are able to learn in order to maintain and adapt themselves. They exert autonomy, but in relation to and as integrative parts of larger systems. They maintain a dynamic balance between structure and flexibility, between order and chaos. In systems terms, these are known as 'complex adaptive systems' and there are no better illustrations than organisms and living systems.

From a systems point of view, the health of any system—be it a family, a community, a farm, a local economy, a school, or an ecosystem—depends on the health of its subsystems, and they on their subsystems and so on. *Sustainability is the ability of a system to sustain itself in relation to its environment*, given that all systems are made up of subsystems and parts of larger supra-systems. A system that either undermines the health of its own subsystems or of its supra-system is unsustainable (see **Box 1** on page 13).

Sustainability is therefore about encouraging self-sustaining abilities and wholeness between systemic levels. It's to do with appreciating and respecting what is already there, with both conserving and developing inherent creative potential, with assisting self-reliance, self-realization, self-sustaining abilities and resilience. From this perspective, it is not difficult to see the parallels between, or the integrative pattern that connects, ecologically sustainable development practice and sustainable education—that connects 'becoming more sustainable' and 'becoming more human'.

Instead of an ethos of manipulation, control, and dependence, the ecological paradigm emphasizes the value of 'capacity building' and innovation, that is, facilitating and nurturing self-organization in the individual and community as a necessary basis for 'systems health' and sustainability. There is a dynamic here which applies differently but similarly to the way sustainability works in say, relation to soil or wildlife management, or developing healthy local economies, or educating children in the classroom. Such principles as diversity, relative autonomy, community, and integrity have an echo in both natural and human contexts. It is only a very short jump to see how educational values such as differentiation, empowerment, self-worth, critical thinking, cooperation, creativity and participation are part of this picture.

In sum, whole systems thinking and ecological sustainability give us bases for envisioning an ecological education paradigm. Following the above model we can now briefly look at *imaging*—picturing how this translates in more detail.

Imaging an ecological education paradigm

The first step is to recognize that all educational thinking and practice takes place within different contextual levels. For example, it is possible to distinguish three interrelated levels where whole systems thinking and sustainability insights can be applied. These are interpreted in **Box 5** in relation to the formal sector, and elaborated in **Table 3** below.

A key idea here is *systemic coherence,* whereby development within and between these three levels is as far as possible mutually reinforcing. In other words, in practice, an ecological paradigm shows both vertical and horizontal integration rather than segregation. These levels of change may be applied at any scale: for example, at international or national level in terms of policy change, or at institutional level, or even at group or individual level. Therefore we can envisage macro, meso, or micro change, all of which are concerned in some way with connection, wholeness and synergy.

Box 5: Three contextual levels to apply whole systems thinking

1. Educational paradigm (Ethos) The implications of ecological thinking as a basis for an overall educational paradigm which revisions and reorients the purpose of education (theory, research and practice) and its relation with wider society and the biosphere. The perceptual dimension—'how do we perceive this'? Key idea: *extension.*

2. Organization and management of learning environment (Eidos) How whole systems ideas might be reflected in systems change and management, organizational ethos, disciplines and departmental structures, curriculum content/theory and design, hidden curriculum, purchasing policy, and community/social links and relationships. The conceptual dimension—'how do we conceive this'? Key idea: *connection.*

3. Learning and pedagogy (Praxis) How whole systems approaches might be reflected in classroom or in community practice, in teaching and learning method, including a systems view of the learner, participative learning and teaching styles. The practice dimension—'how do we do this'?. Key idea: *integration.*

We are now in a position to 'image' an ecological education paradigm in more detail, and at the same time use this to contrast with a model of the dominant mechanistic view. See **Table 3** on pp 58-9, where the three contextual levels from **Box 5** are revisited. Consider: which paradigm are you or your institution serving?

Such an image provides a basis for discussion and *design* of education and learning at any level of operation, and this is looked at again in more detail in Chapter 5. It also gives a basis for thinking through more detailed qualitative indicators, through which progress towards or away from sustainable education may be evaluated. This is an area which needs far more work, not least as reliable indicators appeal to policy-makers.

Outlining such bases for a more ecological paradigm is important, but we need to recognize that progress towards its realization itself involves a learning process, which may not be easy for many.

The learning response of educational systems to sustainability

The challenge of sustainability, as noted in the Introduction, may be viewed as a major threat/opportunity to existing cultural systems and their education systems, and one that requires a paradigmatic rather than piecemeal response. Any positive response involves learning, but the deeper the change apparently required, the more resistance there is likely to be. If we look at the learning response of society as a whole to sustainability, or of educational systems, it varies from nil (i.e. ignorance or denial) through to transformation, which is a deep response. The same range applies to the response of individuals within these systems.

As reviewed in Chapter 1, there are many movements of education for change, which may be seen as struggles towards realizing a more humanistic and ecological paradigm. It is easy to criticize the shortcomings of these movements, but they are mapping paths of transition in often hostile territory. Interestingly, however, they are increasingly converging and agreeing on the importance of transformative learning. Transformative learning "changes who we are by changing our ability to participate, to belong, to negotiate meaning".[8]

The varieties of response by educational thinking and practice to the challenge of sustainability is suggested in the following model (**Box 6**—see pages 60-61) which I've developed from one I first used in the MSc in Environment and Development Education course at South Bank University (see next chapter). These progressive responses, from *accommodation,* through *reformation* to *transformation* may be made at any level—by an individual educator, an institution, or a whole educational system.

The three forms of response seen in **Box 6** can be seen as successive stages of learning in an educational transition paralleling and supporting steps towards sustainability in wider society (this is examined further in chapter 5). Being realistic, the first stage response—'education about sustainability'—is the most likely response in most institutions and countries, and whilst better than no response, it is also the least effective in taking us closer to sustainable living. Critically important in leading and inspiring deeper change are ideas and practices that are working to realize the transition towards sustainable education, and some of these are outlined in the next chapter.

Table 3: Summarizing the contrasting paradigms

MECHANISTIC VIEW	ECOLOGICAL VIEW
LEVEL 1: EDUCATIONAL PARADIGM	
Core Values	
Preparation for economic life	Participation in all dimensions of the sustainability transition—social, economic, environmental
Selection or exclusion	Inclusion and valuing of all people
Formal education	Learning throughout life
Knowing as instrumental value	Being/becoming (intrinsic/instrumental values)
Competition	Cooperation, collaboration
Specialization	Integrative understanding
Socialization, integrating to fit	Autonomy-in-relation
Developing institutional profiles	Developing learning communities
Effective learning	Transformative learning
Standardization	Diversity with coherence
Accountability	Responsibility
Faith in 'the system'	Faith in people
Modernity	Ecological sustainability
LEVEL 2: ORGANIZATION AND MANAGEMENT OF THE LEARNING ENVIRONMENT	
Curriculum	
Prescription	Negotiation and consent
Detailed and largely closed	Indicative, open, responsive
Discursive knowledge	Non-discursive knowledge also valued
Decontextualized & abstract knowledge	More emphasis on local, personal, applied and first-hand knowledge
Fixed knowledge and 'truth'	Provisional knowledge recognizing uncertainty and approximation
Confusion of 'data', 'information' and 'knowledge'	Ultimate concern with wisdom
Disciplines and defence of borders	Greater transdisciplinarity/domains of interest
Specialism	Generalism and flexibility
Evaluation and assessment	
External inspection	Self-evaluation, plus critical support
External indicators, narrowly prescribed	Self-generated indicators, broadly drawn
Quantitative measures	Qualitative as well as quantitative measures
Management	
Synergies & emergence not considered	Positive synergies sought
Architecture, energy and resource use, and institutional grounds neither managed ecologically nor seen as part of the educational experience	Ecological management, linked to educational curriculum and experience

MECHANISTIC VIEW	ECOLOGICAL VIEW
Management (cont.)	
Scale not considered	Human-scale structures and learning situations
Curriculum control and prescription	Curriculum empowerment and determination
Top-down control	Democratic and participative
Community	
Few or nominal links	Fuzzy borders: local community increasingly part of the learning community

LEVEL 3: LEARNING AND PEDAGOGY

MECHANISTIC VIEW	ECOLOGICAL VIEW
View of teaching and learning	
Transmission	Transformation
Product oriented	Process, development and action oriented
Emphasis on teaching	Integrative view: teachers also learners, learners also teachers
Functional competence	Functional, critical and creative competencies valued
View of learner	
As a cognitive being	As a whole person with full range of needs and capacities
Deficiency model	Existing knowledge, beliefs and feelings valued
Learners largely undifferentiated	Differentiated needs recognized
Valuing intellect	Intellect, intuition, and capability valued
Logical and linguistic intelligence	Multiple intelligences
Teachers as technicians	Teachers as reflective practitioners and change agents
Learners as individuals	Groups, organizations and communities also learn
Teaching and learning styles	
Cognitive experience	Also affective, spiritual, manual and physical experience
Passive instruction	Active learning styles
Non-critical inquiry	Critical and creative inquiry
Analytical and individual inquiry	Appreciative and cooperative inquiry
Restricted range of methods	Wide range of methods and tools
View of learning	
Simple learning (first order)	Also critical and epistemic (second/third order)
Non-reflexive, causal	Reflexive, iterative
Meaning is given	Meaning is constructed and negotiated
Needs to be effective	Needs to be meaningful first
No sense of emergence in the learning environment/system	Strong sense of emergence in the learning environment/system

Box 6: Range of educational responses to sustainability

Education about sustainability—This has a content/knowledge bias and can be assimilated quite easily within the existing educational paradigm. This accommodatory response is perhaps exemplified by the newly revised English national curriculum, which takes on board some sustainability concepts. There is an assumption amongst curriculum writers that we know clearly what sustainability is about, that it is uncontested, and that this can be codified and transmitted. Sustainability may be contained as a separate curriculum subject. This is essentially 'learning as maintenance' of the current paradigm because the latter is unchallenged. This is an adaptive response, which equates with first order learning.

Education for sustainability—This includes content, but goes further to include a values and capability bias. This involves some reformation of the existing paradigm to reflect more thoroughly the ideas of sustainability, but otherwise the existing paradigm—even where contradictions are present in espoused or hidden values, for example with respect to unqualified economic growth—remains largely intact. But the emphasis here is 'learning for change', and it is a position that many in the environmental education field advocate. The greening of schools and colleges movement is largely located here. There is often an assumption that we know clearly what values, knowledge and skills 'are needed' but this response does include critical and reflective thinking. This is an adaptive response that equates with second order learning.

Box 6 (continued)

Education as sustainability—This is a transformative, epistemic, learning response by the educational paradigm, which is then increasingly able to facilitate a transformative learning experience. This position subsumes the first two responses, but emphasizes process and the quality of learning, which is seen as an essentially creative, reflexive and participative process. Knowing is seen as approximate, relational and provisional, and learning is continual exploration through practice. The shift here is towards 'learning *as* change' which engages the whole person and the whole learning institution. There is a keen sense of emergence and ability to work with ambiguity and uncertainty. Space and time are valued, to allow creativity, imagination, and cooperative learning to flourish.

In this dynamic state, the process of sustainable development or sustainable living is essentially one of learning, while the context of learning is essentially that of sustainability. This response is the most difficult to achieve, particularly at institutional level, as it is most in conflict with existing structures, values and methodologies, and cannot be imposed. This is third order learning and change—a creative and paradigmatic response to sustainability. In my experience, the best example of this response is probably Schumacher College in Devon.

Chapter Four

Re-making Education and Learning

"The four S's of humanity's survival—Sustainability, Sufficiency, Subsidiarity, and Self-organization—should have a place in everyday knowledge, just like the traditional three R's."—Bossel[1]

A growing movement

If the sustainability transition involves "immense and fundamental changes"[2] in society, then education and learning—so often identified as key agents of change—need to achieve a change of parallel scope and extent. We need to read this as an opportunity for positive change, rather than be daunted by the task.

This brief chapter reviews some initiatives, developments and models that give an idea of what is possible in this field. Despite an educational climate which as yet is largely uncomprehending and unsupportive of such work, initiatives are increasingly coming to light which in some way are trying to embody a more ecological, sustainable and humanistic educational response to our times. As Apple, a radical academic, has said of the American experience: "Not only are large numbers of individuals not 'crushed' by what is happening, but they have rededicated—and very successfully—their lives to building and defending a socially just and caring education worthy of the name."[3]

It is difficult to give any accurate picture of the extent of this response. Nevertheless, it can be said with confidence that there is a growing movement amongst concerned educators the world over, driven both by frustration with mechanistic education and by awareness of the real needs of people and planet.

Thirty years ago, environmental education was a new and virtually unknown term. Now at least it is in the terminology of most governments' educational policy. The take-up of the language and concepts of education for sustainable development is happening more quickly still, even if its paradigmatic implications are only partly understood. In the struggle to embed a broader concept of sustainable education more firmly and adequately, it is the role of the pioneering projects to map out the new ground, to inspire others, and show leadership.

Precursors

Sustainable education is not emerging from a vacuum, but is a potential confluence of many streams of thinking and practice. Some of these influences can be traced a long way back in educational history including humanistic education, progressivism, holistic education, and liberatory education. The contributions of key figures such as Montessori, Steiner, Dewey, Rogers, and Freire need to be acknowledged. The confluence has not yet quite happened, but it may—perhaps under the label of 'sustainable education', or perhaps under something else not yet articulated. Whatever may happen, there is a changing mood towards a more integrative conception of education.

Learning leaders

In the space available, it is not possible to do more than indicate a very few examples of work. The case studies are those that I know of, and are not the result of systematic research. Those interested in further case studies or more detail may view new sources, [4, 5, 6] or check the websites in Appendix II. I've tried to give examples of initiatives and change at *different levels* of educational systems—not least because such change is needed at all levels.

1. Governments working for change

Baltic Sea Region Agenda 21

Getting genuine policy change at national level is hard enough, but at international level it is even more unusual. An initiative in the Baltic Sea area, however, looks promising! In 1996, eleven governments in this area agreed to draw up an Agenda 21 programme with a view to achieving sustainable development in the region. An action programme was agreed in 1998, including a component aimed at strengthening public education and increasing public knowledge of sustainable development in the region. At a subsequent meeting at the Haga Palace in Stockholm in 2000, the

Ministers of Education of the region agreed to develop an Agenda 21 Education Programme, and a network of ministries, authorities and institutions. Working groups are now looking at formal education, higher education, and non-formal adult education respectively, and the first task has been to conduct surveys of existing practice and provision. It is hoped that the governments involved will agree policy reform in the autumn of 2001. Part of the Haga Declaration, which outlines the governments' view of education for sustainable development (ESD), <www.ee/Baltic21/>, is summarized below:

- ESD should be pursued at all levels of education; it should be included in all curricula or equivalent instruments corresponding to the level of education. Such education should rest on a broad scientific knowledge and be both integrated into existing disciplines and developed as a special competence. It demands an educational culture directed towards a more integrative process-oriented and dynamic mode emphasizing the importance of critical thinking, and of social learning and a democratic process.

- ESD should be based on an integrated approach to economic, environmental and societal development and encompass a broad range of related issues such as democracy, gender equity and human rights.

- ESD should also be regarded as an important tool for achieving sustainable consumption and production patterns as well as for necessary lifestyle changes.

- All educational institutions have an important role in the further implementation of Agenda 21 and should aim at: linking to internationally or nationally recognized development strategies or the equivalent; having staff fully trained and competent in ESD; and providing all students with relevant opportunities and methods for learning about sustainable development.

It would be good to see other governments follow this lead!

The Sustainable Development Education Panel

In England, a government advisory Panel has been in operation since 1998. Whilst lacking real power and resources, the Panel has raised awareness of the issues in a number of sectors and some positive change is attributable to its work and influence. Wisely, and from the outset, the Panel has recognized the need for change towards ESD across all sectors of education and society and has developed broad recommendations accordingly.

2. NGOs working for change

NGOs have often led the way in the ESD field. These few examples show how NGOs are working at international, national and local level.

Education 21

A few years after the Earth Summit of 1992, the Education Task Group of UNED Forum, an environment and development organization, launched an integrative concept and rallying cry called 'Education 21' which seeks to involve all members of the 'education community' in implementing the Summit's Agenda 21 programme. It defines the education community very widely, to include all those working in all sectors who have some involvement in formal, non-formal and informal education. This community 'represents an enormously potent, but largely untapped human resource for sustainable development'. The task to galvanize and reorient education as a whole goes on and some efforts are currently being made to get the education community recognized as a 'major group' at the Earth Summit II to be held in 2002.

In Scotland, an alliance of over 50 organizations representing civic society, business and the voluntary sector, is campaigning to make sustainable development a national priority in Scottish educational policy, and the development of the 'sustainable citizen' an entitlement for every child. The Education 21 Scotland alliance is 'committed to education consistent with a sustainable future'.

Learning for a Sustainable Future (LSF)

LSF was established in 1991, with a mandate to integrate the concept and principles of sustainable development into the formal education systems in Canada. Its key educational objectives are that people should develop:

- an understanding of the interdependence between environment, economy and social issues;

- skills in systems thinking, consensus building and decision-making;
- the ability to identify unsustainable practices, find the causes and plan the solutions.

Its main work has been through publication, training and influencing the curricula of the Ministries of Education in the Canadian provinces.

Crystal Waters College

"We strongly believe that students learn by doing a thing, not just hearing about it. If we want to get them excited by a subject we need to let them experience why it's worth knowing."—Morag Gamble of Sustainable Futures.

Crystal Waters, established in 1989, in Queensland, Australia, is one of the world's leading ecovillages and the first to be designed using the principles of permaculture. The Crystal Waters College is a new initiative of 'Sustainable Futures' and the Global Ecovillage Network (GEN) Oceania/Asia centre, both of which are based at the village. A broad range of holistic practical environmental educational programs are offered for people of all ages and back grounds and the College programs attract students from around the world. Sustainable Futures offers an international ecological studies programme, Permaculture Design Certificate Courses and weekend workshops, urban sustainability courses, permaculture camps for high school students, Eco-Village and Healthy House tours—whilst GEN offers courses on permaculture design and cultural exchange to university groups from Japan and North America.

The whole experience is designed to be educational. The entire property of 640 acres is used as a classroom, and a proportion of class time is outdoors doing hands-on activities. In this way, course members directly experience many of the permaculture solutions discussed throughout the education courses. These solutions include: productive permaculture gardens and food forests; biological restoration of soil; integrated animal systems; woodlot establishment; habitat restoration; low energy and healthy houses; sustainable human settlement design; appropriate technology; co-generation; water collection and recycling; a large collection of compost toilet types; active reduction of consumption and waste; community revitalization; co-operatives; and strengthening of local economic systems.

3. Changing schools

I've often heard the cry that teachers and schools 'can do very little' given the pressures and constraints they are under. For some, this is all too true, but very often at least some work towards sustainability is possible: some small change, which can lead to more, and that can empower by its success. At WWF-UK, we found this response quite common, but also that some schools working over a period of time were able to achieve a deeper culture change around sustainability and global citizenship. Now, a small but increasing number of schools are exploring the rich potential of the idea of the 'sustainable school' through an on-going learning process.

Crispin School

In the small Somerset town of Street, in the UK, a state secondary school is continually working towards sustainable education. Through their innovative work, the school has won 'Beacon' status from the government, 'Ecoschool' status from the Going for Green organization, and a curriculum award from WWF-UK for work on education for sustainable development. The school states, "The governing body and senior management team want to make it clear that Crispin supports education for a sustainable future. Of course, this means making sure all pupils have a good understanding of the physical and human environment, so that they know what will lead to a sustainable future and what won't. It also means having a curriculum that develops positive attitudes towards the environment and develops the personal skills that are needed to make informed choices."

One aspect of the school's leading thinking and practice is the development of its 'school's aims' through a whole-school participative process. Broadly, the draft aims are as follows:

"Crispin School believes that pupils should:

- Have a positive view of themselves and the purpose of education;
- Develop the skills and attitudes that are needed to learn effectively and think critically, and value these abilities;
- Be equipped to lead their own lives to the full now and in the future; and
- Be equipped to contribute to a sustainable future."

Box 7: Sustaining change in schools—five key success factors

Over a period of several years, WWF-UK ran a programme called **The Curriculum Management Award Scheme** which sought to embed education for sustainability into the curriculum, ethos and management of a number of schools, through training, funding and action research. Although the scheme included very different kinds of schools, certain common factors emerged which contributed to the achievement of real change. In summary, these are:

Raising staff awareness of sustainability issues: The strands are often already there in ethos and curriculum but not made explicit. The key task is to help staff and pupils fit the pieces together in order to come to a greater understanding of the whole, and from that basis, take it further.

Taking a whole school approach: ESD must be built into policies and schemes of work if it is to go beyond being the personal interest of individual teachers and to survive staff changes. This can only happen with the active support of senior management.

Involving pupils in decision-making processes: There is a need to give pupils the tools needed to participate effectively in the processes of decision-making (individual and collective) and these skills need to be practised like any others. Schools where pupils feel valued and listened to report an improved ethos which enhances learning.

Increasing involvement with the broader community: ESD needs to relate to real situations, local and global. Schools have found a process of opening up to the community very rewarding—providing expertise, resources and a real context for learning, as well as increased mutual understanding and goodwill. Young people can gain confidence and a belief that they can make a difference, and their efforts can stimulate action by parents and the broader community.

Taking one step at a time: Any meaningful whole school initiative is a long term process. Even in schools where staff and governors are fully committed to introducing ESD, small, measurable steps which are monitored and evaluated before moving to the next stage have far more chance of succeeding. It may start with a small nucleus of people, which can slowly begin to influence a year group or department.

Gillian Symons, WWF-UK consultant

These aims have been elaborated in detail with versions both for pupils and for staff.

The Small School

"There is time to do the unusual, and a freedom to express things."— Small School student

The Small School in Hartland, Devon is one of the leaders in exploring the meaning and implications of sustainable education. A member of the Human Scale Education (HSE) movement, the school has worked to develop a working humanistic and ecological philosophy and methodology.

Caroline Walker, former head teacher, writes: "The Small School follows Schumacher's dictum: 'It is of very little use to call for big changes in values without incorporating those values in some new structures, no matter how small.' "

The Small School sets out to foster community, not individualism; simplicity rather than consumerism; spirituality rather than materialism. Smallness allows participation, positive relationships, and whole-school policies for environmental and social sustainability: an egalitarian pay structure, an ethical purchasing policy favouring organic, fair trade, local and recycled goods, a curriculum that includes practical skills like growing food and cooking, knowledge of global issues like climate change, fair trade and world debt, and values of spirituality through silence and a daily Peace Prayer. Natural materials are used in classrooms, and the grounds are managed for wildlife, without the use of chemicals. Founded by a group of parents and teachers in 1982, the school has a maximum of 40 students taught by qualified teachers and local craftspeople. Old students can be found in a wide variety of occupations from agricultural work to university lecturing.

The Small School and the wider HSE movement work for change through the promotion of the principles and practices of human scale education in mainstream schools as well as offering an alternative model of schooling.

4. Changing teacher education

The world has around 60 million teachers, very few of whom have had any exposure in their training to sustainability issues or to the ideas of sustainable education. Even where sustainability is in the curriculum in some form, and even where there is personal interest in the area, the majority of teachers have little idea about how to teach this area, or about its deeper pedagogical and other implications. The provision of in-service training is one immediate solution, but in the medium term it is more important to reorient pre-service training. There is still a great deal to do in this area. The case study below is of a unique research project which has recently been put on-line by UNESCO, and this may stimulate further change.

Learning for a Sustainable Environment[7]

One major programme that started addressing this problem is the 'Learning for a Sustainable Environment: Innovations in Teacher Education Project', which has worked with teacher educators from nearly 30 countries in the South East Asia and Pacific region. Led by the UNESCO Asia-Pacific Centre for Educational Innovation for Development, Bangkok, and Griffith University, Brisbane, the project had two goals:

- to develop a model for providing effective professional development in the knowledge, skills, and values of environmental education for teacher educators throughout the Asia-Pacific region.
- to provide carefully trialled and culturally sensitive workshop materials that could be used as the basis of professional and curriculum development activities by teacher educators in other countries and in other parts of the world.

Using an action research network model of curriculum change, the project also served as a professional development process for teacher educators. The collaborative process encouraged the participants to experiment with, and incorporate environmental education knowledge and skills into their own teacher education courses and programmes. The preservice and inservice teachers in the classes of network participants received current and relevant knowledge and skills to introduce or improve environmental education in their own classrooms. Thus the project updated both teacher education practices and practitioners throughout the region.

The revised modules are now available on the Internet. A development from this project is the UNESCO multimedia teacher education programme 'Teaching and Learning for a Sustainable Future'.

5. Changing higher education

In responding to sustainability, higher education has been one of the slowest sectors to take up the challenge. The comparative newness of the sustainability agenda, the interdisciplinary and transdisciplinary nature of the area, the need for learner-centred approaches, 'green management' and organizational learning, all pose a challenge to established norms. However, there are some encouraging signs of change in several countries. In the UK, there are now funded pilot projects that are beginning to address these issues, both in higher and further education; for example the Higher Education Partnership coordinated by the NGO Forum for the Future, as part of its Education and Learning Programme. The role of innovative exemplars in moving HE forward is vital, and a few brief case studies follow.

Schumacher College, Devon

"I had been told that the College provided a unique, wonderful experience, but I did not anticipate that it would be such a refined blend of intellectual, spiritual, physical and social experiences."—A course participant

Schumacher College, an 'international college for ecological studies' has won a pioneering reputation. Through courses covering such areas as ecological economics, ecological design, community, spirituality and the arts, as well as holistic science, the college seeks to explore the foundations and implications of ecological thinking and practice. Through paying close attention to the learning environment, to the needs of the whole person, and in fostering learning communities, the college seeks to respond to the challenge of being an engaged and transformative institution. Many other institutions internationally have been influenced by it. The College has also recently introduced short courses which give business practitioners the opportunity to look at the shift required in both values and technology if sustainability is to be truly integrated into business.

Schumacher College has also broken new ground in offering a MSc in Holistic Science in partnership with the University of Plymouth. The course 'goes beyond reductionism in understanding dynamic interactive systems', cultivates intuitive as well as analytic ways of knowing, and explores a 'science of qualities'.

Legitimizing innovation—the Masters route

"The course by no means provides me with all the answers, rather it encourages me to produce yet more questions and to find the answers to those out in the real world."—SBU distance learning student working on community based education and agriculture, Indonesia.

Other institutions are increasingly using the same strategy: offering an innovative Masters level course which gives credibility to the area, supports new research and leads to dissemination as graduates take their expertise and thinking elsewhere. An example is the MSc in Environment and Development Education at South Bank University (SBU) in London, which has pioneered the theory and practice of education for sustainability through full-time and distance learning courses that have attracted students from around the world. Students are experienced practitioners in schools, colleges, community education, NGOs, government bodies and development agencies, all of whom develop and apply sustainable education ideas in many different contexts and countries, and try to close the gap between theory and practice. The course has directly given rise to similar initiatives in Hungary and China.

The MSc in Responsibility and Business Practice, at the Centre for Action Research in Professional Practice at the University of Bath, UK, is another example of innovation in the area of education for sustainability at Masters level. The course 'addresses the connections between the management of business activities and the major social, political and environmental issues of our time' and is 'concerned with developing a new management philosophy where social, ethical and environmental criteria are central to business policy, alongside financial issues'.

'The pedagogy of architecture'

A problem for sustainability courses is that the buildings and environment in which they are held often contradict the ideas being studied. Leading environmental educator David Orr has pioneered the development and construction of a new teaching facility at Oberlin College, Ohio, USA, which is itself a teaching tool.

The ecologically-designed Adam Joseph Lewis Environmental Studies Center represents what Orr has called 'the pedagogy of architecture', that is, the curriculum is embedded in buildings in terms of the materials used, the design, and its ecological footprint. The Center includes a Living Machine organic waste system that purifies and recycles waste water, photovoltaic roof cells that produce electricity, and compostable, non-toxic and recycled materials. The Center aims for zero-emissions. The whole process of designing the building was democratic and had educational value, involving staff, students and the local community. Oberlin hopes that the building will stimulate a new direction in the campus-greening movement.

For more on changing the campus and curriculum, see the 'Second Nature' and 'IAU' websites in Appendix II.

6. Changing business and professional practice

In only a very few cases have organizations begun to address the professional development of their staff, focussing on learning, their personal values, as well as a review of business practice, purpose, vision and values.
Professional Practice for Sustainable Development project (PP4SD)[8]

In this area, the key issue is closing the gap between business interest in sustainability and what happens in most business training, particularly in management. In the UK, a number of initiatives are now underway which seek to introduce sustainability thinking into the vital area of professional training and practice. One example is The Natural Step (TNS) programme which offers training courses particularly to the corporate sector based on the 'TNS Framework'. This is a systems-based planning methodology that stipulates four 'system conditions' through which organizations can work towards sustainability (and there is also a schools project based on this approach). Another initiative is the PP4SD project, which is run by the Institution of Environmental Sciences, in association with TNS, the Environment Agency, WWF-UK and the Council for Environmental Education. This project focuses on inter-professional learning and has developed an accessible framework designed to encourage and support the integration of sustainable development (SD) principles into the practice of professionals such as engineers, waste managers, architects, and planners. Currently, fourteen professional institutions are taking part in the project. Two booklets and an SD foundation course have been produced, and training courses are planned.

Local authority change

A number of local authorities in the UK are moving on from Local Agenda 21 strategies, and recognizing that change towards sustainable communities hinges on learning and education. To that end, posts with 'education for sustainable development' in the title are beginning to appear. In Dorset, UK, for example, a partnership between WWF-UK and Dorset County Council is funding a study and co-ordinator to promote contact between stakeholders interested in sustainable development, and promote networking between educators and their service providers. In Worcester, UK, an ESD strategy has been produced which aims to work across the different county council directorates, services and agencies, and ultimately provide ESD for all sections of the community. For other examples of the work of local authorities on sustainability, see <www.la21-uk.org.uk> and the International Council for Local Environmental Initiatives (ICLEI) website <www.iclei.org>.

What's important?

Space does not permit discussion of curriculum here, but to give some stimulus, I have included summaries of a number of schema around which sustainable education might focus. There is increasing work on the curriculum and pedagogic implications of sustainability, and many more schema and ideas may be found on ESD websites (see Appendix II). Whatever curriculum is agreed, detailed and restrictive 'learning outcomes' are questionable. Far better to develop—in a participative way, of course!—general, indicative concepts which are then explored, critiqued, adapted and made relevant within local and regional contexts.

Pillars of life-long learning

In 1996, an international commission report to UNESCO proposed four pillars as the foundations of education.[9] These are in sum:

- learning to live together;
- learning to know;
- learning to do; and
- learning to be.

The on-going debate on the implementation of these principles in education can be found on the UNESCO website <www.unesco.org>.

Global citizens

A survey of experts in nine countries, East and West, revealed a consensus on eight characteristics that will be needed by citizens of the 21st century to cope with, and constructively engage with, likely major global trends. They are ranked as follows:[10]

- looking at problems in a global context;
- working co-operatively and responsibly;
- accepting cultural differences;
- thinking in a critical and systemic way;
- solving conflicts non-violently;
- changing lifestyles to protect the environment;
- defending human rights; and
- participating in politics.

These may be seen as guiding principles in any curriculum for what is increasingly being called 'global citizenship'.

Sustainable development key concepts

In 1998, I chaired a working group that produced a paper on education for sustainable development (ESD) for the Government Sustainable Development Education Panel in England. In this paper we suggested seven key concepts of sustainable development that should be covered and critiqued in curricula, and then suggested how these would break down into general learning outcomes for different ages. They were as follows:[11]

• Interdependence—of society, economy and the natural environment, from local to global;

• Citizenship and stewardship—rights and responsibilities, participation, and co-operation;

• Needs and rights of future generations;

• Diversity—cultural, social, economic and biological;

• Quality of life, equity and justice;

• Sustainable change—development and carrying capacity; and

• Uncertainty, and precaution in action.

Components of ecoliteracy

The Center for Ecoliteracy in California, founded by Fritjof Capra, aims to promote learning among the young to help them build ecologically sustainable communities. The Center's work with schools in California is based on a number of principles:[12]

• Understanding the principles of ecology, experiencing them in nature, and thereby acquiring a sense of place. These principles are: networks, nested systems, cycles, flows, development/co-evolution, and dynamic balance;

• Incorporating the insights from the new understanding of learning, which emphasizes the child's search for patterns and meaning;

• Implementing the principles of ecology to nurture the learning community, facilitating emergence, and sharing leadership; and

• Integrating the curriculum through environmental project-based learning.

Chapter Five

Reorienting Education—Designing for Change

"A sustainability revolution requires each person to act as a learning leader at some level, from family to community to nation to the world."—Meadows[1]

The need for learning leaders

This Briefing has reviewed the meaning of sustainable education, its bases and some of its implications, it has also addressed the issues of educational change and the nature of educational paradigms. Some evidence of progress towards sustainable education has also been examined. However, the re-direction of education as a whole towards sustainability remains a serious challenge.

As education for change is by definition a long-term process, the time available to transform education so that it might itself be transformative is diminishing, when seen against global unsustainable trends in areas such as resource use, population, poverty, inequity and loss of environmental diversity and quality. Moreover, mainstream education is slow to change, and has a tendency towards conservatism. Internationally, since the 1996 Delors' report on 'Education For the 21st Century', which favoured an holistic and humanistic emphasis in education, this orientation has largely been ignored in favour of managerialism. But many educators are realizing that this latter paradigm is unsustainable. This chapter sets out some frameworks relating to the 'ecological design' of education that might help those working for change towards sustainability, whether from within or from outside 'the system'.

Progressive change in the sustainability transition

How do we transform education? At micro level, in for example, the exchange you next have with a child or adult might be transformative immediately. You might effect "a movement of mind".[2] In human-scale learning situations, the potential for transformative learning is always more immediate. In my own experience, working with teachers with WWF-UK in-service programmes, it has often been possible to see a real change, a real awakening, deep reflection and revitalization of enthusiasm take place. Our 'graduates' often rediscovered 'what they had gone into teaching for', felt more able to reconcile their beliefs with what they actually did or were required to do, were keen to work cooperatively, were more able to be creative and innovate, and be reflective 'change agents'. At the macro level of the institution or organization, however, transformative and systemic change is much more difficult.

At individual *or* institutional level, the learning response to sustainability may be non-existent, or superficial, or progress through a spectrum of change. Possible learning stages, as outlined in Chapter 3, may be:

- *No change* (no learning: ignorance, denial, or tokenism);
- *Accommodation* (first order learning, adaptation and maintenance);
- *Reformation* (second order learning, critically reflective adaptation); or
- *Transformation* (third order learning, creative re-visioning).

Interestingly, there is a parallel between progressive change *in education,* as suggested here, and change *in society* towards what a number of writers call 'strong sustainability'. In a major study of the progress towards the implementation of Agenda 21 in Europe, a similar four-stage model of sequential change was distinguished in four areas of interest: economic policy, environmental policy, public awareness and public discourse.[3] A shift towards the transformative, strong sustainability end of the spectrum would mean increasing integration between these four areas of interest, greater subsidiarity and greater sustainability, with the last stage involving:

- a cultural shift in education and public awareness;
- much closer integration between environmental and economic policy; and
- a renewal of emphasis on local democracy and activity.

It is important to notice that the progression that any of the four areas of interest makes through these stages needs to occur roughly at the same time as the others. In other words, it is not possible for 'education and public awareness' to move far ahead of the other shifts. However, at present there is not much danger of that: formal education is largely lagging behind in this sustainability transition. Therefore those interested in working towards sustainable education need to tune into the broader debates and movements that now characterize the sustainability field—it is difficult to design for sustainable education if we have little idea of current progress towards making a more sustainable society.

In those educational systems and institutions where there has been no response at all to sustainability, a necessary first step is to work for 'education about sustainability' as an add-on to existing policy and practice. This is an *accommodatory* response, which some systems have adopted, but it is insufficient. Beyond this, the next stage to tackle is the *reformatory* stage. This is more difficult and what is often meant by 'reorientation'. But at the same time, we need to maintain a *transformatory* vision so that change does not become stuck at the early levels of change, corresponding with weaker sustainability, but is able to work through to deeper levels.

A vision for change . . .

Transformation depends on a vision informed by a clear philosophy. This again brings us back to the *purpose* of education. The commonly held purpose of any system, for example a business or educational institution, is key to its functioning, which is why strategy-building processes always try to clarify the purpose or mission of an organization. Change agreement around that and everything else tends to change. If governments changed their conception of education and announced its primary role was to contribute to a sustainable society, it would be momentous. But after some 25 years' involvement in lobbying government on environmental education and education for sustainability, I think this unlikely.

Outside certain safe parameters, governments and most policy makers at all levels tend to follow rather than lead. Furthermore, many leaders "have lost both the habit of learning and the freedom to learn".[4] So it is vital that all those concerned about education should discuss, debate and develop some vision of sustainable education. A changing climate of debate and practice is necessary to effect a real change in policy climate, and reciprocally, a change in policy can only be effective where conditions are fertile for change. In other words, to use Gandhi's phrase, "We have to be the change we want to see".

But we also need to articulate it. By giving it expression, it is more likely to be created and emulated. I believe 'unsustainable education' will have to change towards sustainable education eventually. There is increasing evidence of positive change, but time is short. What I am interested in is how this process can be consciously accelerated. I've spoken to people who believe that education systems are in such a stage of criticality that they are ready to 'breakthrough' to a new paradigm—if it is sufficiently clear what it means and implies. This raises the issues of design and action by concerned educators at all system levels to explore, articulate and give expression to sustainable education as a healthy and meaningful alternative.

. . . and of change

It is useful to make a distinction here between strategic *planning* (which tends to work from the existing position and prescribe goals) and ecological *design* (which looks at a vision of what is needed, and explores the values and systems that can work towards its realization on a basis of on-going learning). *Planning and targets* tend to be mechanistic, controlling, specific and time-bound; *design and aspirations* tend to be organic, participative, open, iterative and evolving. From the current situation of overplanning, and too much top-down control—which leads to rigidity and inability to respond to change—we need to move towards the 'sustainable design' of education.

Considered at any level of educational systems, there are two key principles:

- the process of change influences the products of change; and
- emergent properties cannot be predicted but can be designed for.

How can we design in an open and non-deterministic way, educational systems and institutions that promote healthy emergence? This is an absolutely critical question, which begs another: can we close the gap between what we feel education should be and do, with what actually happens, through intelligent and reflective design and action? This depends partly on developing a model of an aspired-to system of sustainable education.

Box 8: The concept of emergence

Emergent properties are hugely important in education, yet are hardly recognized. The idea of emergence is at the heart of the new sciences of complexity. Fundamentally, it describes the qualitative properties that arise from the interaction of parts or individuals in a complex system, and which are not reducible to, or predictable from, these parts.

In the context of organizations, including educational institutions, emergence describes the 'living qualities' that arise and change from dynamic interaction. These are critical to educational organizations and process at all levels of experience. For example: the degree and types of trust, or stress, or innovation; the quality of learning both intended and unintended; the spirit or atmosphere in the organization or classroom; the feel of the learning environment; the overall 'health' of an institution or community; and the additionality that arises from transdisciplinarity.

Whilst emergent properties cannot be controlled, let alone measured, we can design learning environments and facilitate for healthy and positive emergent properties. This will give us increasing 'win-win' synergies. Good teachers and leaders appreciate emergence intuitively and know how to work with it constructively. But under mechanistic educational regimes where emergence is not recognized, the emergent qualities that nevertheless arise can often be negative. With emergence in mind, the idea of quality in education takes on a whole different meaning and relates to the total experience.

Going forward from the *vision* (underlying philosophy), and *image* (picture of core values and ideas) of such a system, as sketched in Chapter 3, we can consider the *design* which is likely to move the institution or system in the desired direction. It is important that design should be seen as a continuous learning process rather than a blueprint.

There should be a participative process of modelling which can be carried forward into actual development, management, and evaluation as an on-going learning process involving all players. This is not a mechanistic exercise, but one that demands imagination and collaboration. Generally, people support change they feel they *own*,

Box 9: Four components of design

Developing:
A statement of purposes: for an education system or institution. What is it for? What should it do? In systems terms, this is a 'root definition'.
The parameters of the system: who owns it, who are the 'clients', what services does it provide, how does the system relate to the community, society or government?
A 'functions model': what does the system need to do to fulfil the purposes? How do the functions relate to each other?
The management and organization: what systems, structures, people, capacities, competencies and actions are likely to be needed to work towards realization of the all the above? (After Banathy[5]).

and resist change they feel is externally *imposed*. Innovation stands more chance of success if it is perceived as 'helpful', 'plausible', 'possible', and likely to be 'fruitful'.[6]

There are a number of points to make about the above model (see **Box 9**).

- *Stating the difference:* envisioning a desired system in this way allows us to perceive more clearly the difference between 'where we are', and 'where we want to be', and ways to bridge the gap.

- *System levels:* this kind of model can be used at all levels of education, from the classroom or lecture theatre, to institutional change, to local and national policy level. Indeed, change towards sustainable education requires a whole system approach at all levels.

- *Balancing the visionary and practicable*: these need to be mutually informing. "Vision without action is useless, action without vision is directionless".[7] What's practicable now needs to be informed by a strong sense of an ecological paradigm for education, whilst this vision needs to be informed by what can be started today.

- *Possibility*: some positive change, however small, is always possible, and may affect the whole in ways that cannot be anticipated and may never be known. This is an empowering idea!

- *Complexity and control*: forget controlling detail and instead, embrace leadership that designs for and nurtures participation, co-creation, healthy emergence, and self-organization.

Brenda Gourley, writing from the experience of attempts to introduce systemic change at the University of Natal, South Africa, states: *"To aspire to being 'a learning organization' it is necessary to build a shared vision, share mental models and learn as a team"*.[8]

Taking it from here

While it is difficult to generalize or make specific recommendations, as this Briefing is written for anybody in any situation in relation to education and learning, some ideas for the short term follow, roughly divided into 'policymakers' and 'practitioners', though some ideas will apply to both.

Suggestions for policymakers and leaders

- Be open-minded and accept you may need to learn much more about sustainability and sustainable education. Realize that 'efficiency' and 'improvement' in education and institutions are not sufficient responses to the challenge and crisis of sustainability.

- Recognize that sustainable education is critical to the realization of 'joined-up' policy in the areas of environment, economy and society and its ability to effect change. This should not just be the concern of education policy makers, but also their colleagues working on policy for commerce, environment, health, development, and social issues etc.

- Consider how to reorient education as a whole towards a more ecological, integrative, and sustainable model through gradual steps, and set up research and monitoring frameworks that can help you facilitate this process. Gather expertise on systems change in relation to sustainability and complexity, but also value existing expertise and commitment on the ground.

- Change the rationale and purposes of education under your influence to make *learning towards sustainable living* an explicit, central and integrating concept in educational planning and practice. Then promote this mandate and sense of common purpose.

Box 10 : One vision of sustainable education

Here is my vision of sustainable education, drawn at very general level. It may not be yours, but I hope it stimulates your own thinking and inquiry. It is an elaboration of the three-part model outlined in Chapter 3. Sustainable education is:

Extended . . .
Appreciative—aware of the uniqueness and potential of each individual and group, of the qualities of any locality and environment, and sees personal and local knowledge as foundational to learning.
Ethical—extends the boundaries of care and concern from the personal and the present, to the social, environmental, non-human, and future dimensions.
Innovative—draws inspiration from new thinking and practice in a variety of fields, relating to education, learning and aspects of sustainable development.
Holistic—relates to the learning needs of the 'whole person' (including spiritual and emotional), of differentiated individuals and groups, and to the range of human intelligence.
Epistemic—aware of its own worldview and value bases, which are critically examined and reviewed. Second and even third order learning is facilitated.
Future oriented—concerned with creating a better future, from now on.
Purposeful—critically nurtures sustainability values with the intention to assist healthy change.

Connective . . .
Contextual—in touch with the real world, particularly sustainability issues, and grounded in the locality.
Re-focused—particularly on social development, human and natural ecology, equity, futures, and practical skills for sustainable living.
Critical—ideologically aware, deconstructive and constructive.
Systemic—pays attention to systemic awareness of relationships, flows, feedbacks, and pattern in the world.

Box 10 (continued)

Relational—connects patterns of change: local-global, past-present-future, personal-social, environmental-economic, human-natural, micro-macro etc.
Pluralistic—values different ways of knowing, and multiple perspectives.
Multi and transdisciplinary—regards disciplinary borders as fuzzy and puts greater emphasis on new ways of seeing complex issues.

Integrative . . .
Process oriented—constructs meaning through an engaged and participative learning process, reflecting different learning styles. Everyone is a learner, including the teacher/leader.
Balancing—embraces cognitive and affective, objective and subjective, material and spiritual, personal and collective, mind and body etc.
Inclusive—for all persons, in all areas of life and extending throughout their lifetimes.
Synergetic—deeply aware of emergence, and designs curriculum, organization and management, culture to be mutually enhancing. Energy, material, and money flows are organized on sustainability principles and are reflected in the whole curriculum.
Open and inquiring—encourages curiosity, imagination, enthusiasm, innovation, creativity, community and spirit to arise. At ease with ambiguity, and uncertainty.
Diverse—allows for variety, innovation and difference of provision and ways of knowing within a coherent framework.
A learning community—the institution promotes learning through itself engaging in reflexive learning (a learning organization).
Self-organizing—balancing autonomy and integration through different system levels and practising subsidiarity and democracy.

Such an education and learning situation would be intrinsically transformational—of itself, and of its community members—and would have systemic coherence.

• Recognize the need to move from emphasis on functional learning towards higher-order learning in education, including much more inclusion of systems thinking and critical thinking.

• Reform policy-making and administration on a basis of trust, inclusivity and subsidiarity, and evaluation on a basis of shared and supported self-evaluation. Say less, listen more. Judge less, support more.

• Consider how new emphases of the quasi-market educational model such as flexible learning patterns, life-long learning, and IT skills can be reoriented and revisioned to reflect the need for skills for sustainable living.

• Recognize that formal education needs to be seen as an integral and contributing part of a broader movement of social learning towards sustainability, including media, family, community, and workplace learning.

• Commit resources and funding, not least for staff development.

• Ease up! Create space, time and a supportive atmosphere where people feel enabled and secure enough to reflect, think, and innovate.

• Retain the democratic control of education and resist the increasing corporate influence on educational policy and priorities.

Educational policies can help to create a better world, by contributing to sustainable human development, mutual understanding among peoples and a renewal of democracy.[9]

Suggestions for practitioners

• Value the people in your group, community, school, college, more than top-down policies. People come first! Recognize your strengths.

• Contemplate and discuss with others what you really feel education and learning to be, and what they are for. What do students really need to learn? Why?

- Where necessary, help policy-makers, colleagues, and the community learn about the meaning of, and necessity for, sustainable education in a way which builds on their own knowledge and values. Keep communication levels high.

- Fight for self-assessment and self-evaluation; for local support and 'critical friending' rather than judgemental inspection.

- Work with policy makers and draw up 'reorientation schema' for use with educational policies, plans, curriculum statements and learning outcomes, indicating with examples, how to distinguish between concepts/goals that should be abandoned, modified, or newly written in order to address sustainability.

- Use the new interest in citizenship as a opening to develop a more holistic and inclusive 'ecological citizenship' that can help transform policy and practice and address issues of equity, human rights and needs, consumerism, community, participative democracy and sustainability.

- Explore systems thinking and use it to understand the system you are working in. Begin to use it to get young people and adults to think critically and connectively.

- Inform yourself as much as possible on sustainability issues, particularly at the local level.

- Find allies in and beyond your work area in terms of people, resources, ideas, using personal contact and the Internet including ESD websites.

- Network with others and help to build local support networks.

- Be inspired by the good things going on. Communicate positive stories and celebrate successful change.

- Create the future: think of and do one achievable thing that will bring sustainable education further into reality, then another . . .

The real choice

The advocates of market-based education are fond of talking about choice. But their notion of choice is a little like choosing deckchairs on the *Titanic.* We have a much bigger and much more historic choice to make: whether or not we will move towards sustainable education, say within five years. At the time of writing, it seems unlikely that education will be addressed at the official Earth Summit 2002 in Johannesburg. All the more reason then for all concerned not to wait, but to take action now.

Sustainability means working to understand and realize sustainability values in ways where economic, social and ecological dimensions are as far as possible mutually enhancing. It's about creating the conditions of survival, security and wellbeing for all. Un-learning, re-learning, new learning are the essences of this challenge. This imperative of ecologically sustainable development gives education a powerful purposeful orientation that challenges the market-led paradigm.

Seen positively, paradigmatic change in education can best proceed from alliance with parallel change towards sustainability in wider society. The alternative is reactive change as the critical environment/development issues bite more severely as the new century progresses, but by then we shall have lost precious time.

A second compelling reason for sustainable education is the crisis in education. If we can develop an inclusive model where everybody feels valued and able to make a difference, the educational benefits, and indeed, social and economic benefits, will be enormous.

In his analysis of conventional economics, Herman Daly[10] calls for a shift from the information economy to the 'wisdom economy'. We have a parallel and related challenge, to shift from 'information education' to 'wisdom education'. Time is short, and we need to summon all our existing wisdom to choose and enact this change.

Appendix I

References

Introduction

1 Schumacher, E. F. (1973) *Small is Beautiful,* Blond and Briggs, London.
2 Source: *The Living Planet Index,* WWF, 1999.
3 Source: *World Education Forum,* Dakar, Senegal 26-28 April 2000.
4 Meadows, D.H., Meadows, D.L., and Randers, J. (1992) *Beyond the Limits: Global Collapse or a Sustainable Future,* Earthscan, London.
5 O'Riordan, T. and Voisey, H. (1998) *The Politics of Agenda 21 in Europe,* Earthscan, London.
6 Laszlo, E. (1997) *3rd Millenium: The Challenge and the Vision,* Report of the Club of Budapest, Gaia Books, Stroud.
7 Clark, M. (1989) *Ariadne's Thread: The Search for New Ways of Thinking,* Macmillan, Basingstoke.
8 Senge, P. (1990) *The Fifth Discipline,* Doubleday Currency, New York.
9 WCED (1987) *Our Common Future,* World Commission on Environment and Development, Oxford University Press, Oxford.

Chapter 1

1 Schumacher, E. F. (1997) *This I Believe and Other Essays,* Green Books, Dartington, UK.
2 Orr, D. (1994) *Earth in Mind: on education, environment and the human prospect,* Island Press, Washington.
3 Kosko, B. (1994) *Fuzzy Thinking,* Flamingo.
4 Reichel-Dolmatoff, G. (1996) *The Forest Within: The World-View of the Tukano Amazonian Indians,* Themis Books, Dartington.
5 Sterling, S. (1996) 'Education in Change' in Huckle, J., and Sterling, S. *Education for Sustainability,* Earthscan, London.
6 Hicks, D. (1994) *Preparing for the Future: Notes and Queries for Concerned Educators,* WWF-UK/Adamantine Press.
7 Fien, J. (2000) 'Listening to the voice of youth: implications for educational reform, in Yencken, D. et al (2000) *Environment, Education and Society in the Asia-Pacific,* Routledge, London.
8 Clark, M. (1989) *Ariadne's Thread: The Search for New Ways of Thinking,* Macmillan, Basingstoke.
9 Banathy, B. (1991) *Systems Design of Education,* Educational Technology Publications, New Jersey.
10 Koestler, A. (1989) *The Ghost in the Machine,* Arkana, London (first published by Hutchinson, 1967).
11 Daly, H. (1996) *Beyond Growth: The Economics of Sustainable Development,* Beacon Press, Massachusetts.

Chapter 2
1 O'Sullivan, E. (1999) *Transformative Learning: Educational Vision for the 21st Century*, OISE/UT/Zed Books, London.
2 Green, A. et al (1999) *Convergence and Divergence in European Education and Training Systems,* Institute of Education, University of London, London.
3 UNESCO, (1978) *Declaration of the First Intergovernmental Conference on Environmental Education*, Tbilisi, 1977.
4 UNCED, (1992) *'Promoting Education, Public Awareness and Training'*, Agenda 21, Chapter 36, UNCED, UNESCO, Paris.
5 Mayor, F. (1997) 'Preface' in UNESCO, *Educating for a Sustainable Future,* UNESCO, Paris.
6 UNESCO Secretary General (2000) *Progress Report on the Implementation of the Work Programme on Education, Public Awareness and Training*, UNESCO, Paris.
7 UNESCO, (2000) *World Education Report 2000*, UNESCO, Paris.
8 Matsuura, K. (2000) 'Hard Lessons for the World's Educators', *Financial Times,* April 26, London.
9 Chambers, R. (1997) *Whose Reality Counts? Putting the First Last,* Intermediate Technology Publications, London.
10 Smyth, J. and Shacklock, G. (1998) *Re-Making Teaching: Ideology, policy and practice,* Routledge, London.
11 Quoted in Barlow, M., 'The Last Frontier', *The Ecologist*, February 2001.
12 Source: *'Schools Embrace Big Business to Raise Money in Canada'*, by Paul Weinburg, Inter Press Service.
13 Bassnett, S. (1998) 'The View from Here', *Education, The Independent*, 12/2/98.
14 Laurillard, D. (1999) 'A conversational framework for individual learning applied to the "learning organization" and the "learning society" ', *Systems Research and Behavioural Science*, vol. 16, no. 2, Wiley.
15 Tate, N. (2000) 'Make mine a A', *Education Guardian, 1/2/2000.*
16 Roszak, T. (1981) *Person/Planet*, Granada, London.
17 Clark, M. (1989) *Ariadne's Thread: The Search for New Ways of Thinking,* Macmillan, Basingstoke.
18 Bayliss, V. (1999) *Redefining the Curriculum,* Royal Society of Arts, London.
19 Bentley, T. (1998) *Learning beyond the classroom: education for a changing world,* Demos/Routledge, London.
20 de Geus, (1997) *The Living Company,* Nicholas Brealey, London.
21 Roberts, P. (1998) 'Challenging the assumptions of change', *Developing People,* no. 3, vol. 11, Roffey Park Management Institute, Horsham.

Chapter 3
1 Elgin, D. (1997) *Global Consciousness Change: Indicators of an Emerging Paradigm*, Millennium Project, California.
2 Capra, F. (1996) *The Web of Life,* HarperCollins, London.
3 Ho, Mae-Wan (1998) *Genetic Engineering: Dream or Nightmare?* Gateway Books, Bath.
4 Korten, D. (1999) *The Post-Corporate World,* Berrett-Koehler, San Francisco.
5 Reason, P. and Bradbury, H. (eds) (2000) *Handbook of Action Research: Participative Practice and Enquiry,* Sage Publications, London.

6 Banathy, B. (1991) *Systems Design of Education,* Educational Technology Publications, New Jersey.
7 Bateson, G. (1980) *Mind and Nature: a Necessary Unity,* Bantam Books, NY.
8 Wenger, E. (1998) *Communities of Practice: Learning, meaning and identity,* Cambridge University Press, Cambridge.

Chapter 4
1 Bossel, H. (1998) *Earth at a Crossroads: Paths to a Sustainable Future,* Cambridge University Press.
2 O'Riordan, T. and Voisey, H. (1998) *The Politics of Agenda 21 in Europe,* Earthscan, London.
3 Apple, M. (1995) *Education and Power,* 2nd edition, Routledge, London.
4 de Haan, et al (eds) (2000), *Educating for Sustainability,* Peter Lang, Frankfurt.
5 Wheeler, K. and Bijur, A. (2000) *Education for a Sustainable Future,* Kluwer Academic/Plenum Publishers, New York.
6 Clover, D., Follen, S. and Hall, B. (2000) *The nature of transformation: environmental adult education,* OISE, University of Toronto.
7 From communication with John Fien (Director of the Centre for Innovation and Research in Environmental Education, Griffith University, Australia).
8 Professional Practice for Sustainable Development (2001) Book 2: *'Developing Cross-Professional Learning Opportunities and Tools',* Inst. of Environmental Sciences.
9 Source: International Commission on Education for the 21st Century, UNESCO.
10 Cogan, J. and Derricott (eds) (2000*) Citizenship for the 21st Century: An International Perspective on Education*, Kogan Page, London.
11 Sterling, S. (ed) (1998) *Education for Sustainable Development in the Schools Sector, A report to the DfEE/QCQ by the Panel for Education for Sustainable Development*, CEE, Reading.
12 Source: Center for Ecoliteracy, 2000

Chapter 5
1 Meadows, D.H., Meadows, D.L., and Randers, J. (1992) *Beyond the Limits: Global Collapse or a Sustainable Future,* Earthscan, London.
2 Senge, P. (1990) *The Fifth Discipline,* Doubleday Currency, New York.
3 O'Riordan, T. and Voisey, H. (1998) *The Politics of Agenda 21 in Europe,* Earthscan, London.
4 As 1 above.
5 Banathy, B. (1991) *Systems Design of Education,* Educational Technology Publications, New Jersey.
6 Fensham, P. (1994) 'Issues Influencing the Adoption of Innovation in Teacher Education', in *Learning for a Sustainable Environment* Final Report of Planning Group, Griffith University/UNESCO-ACIED, Brisbane.
7 As 1 above.
8 Gourley, B. (1999) 'Systems and Transformation at the University of Natal', *Systems Research and Behavioural Science,* vol. 16, no. 2, Wiley.
9 Delors, J. (chair) (1996) *Learning: The Treasure Within: Report to Unesco of the International Commission on Education for the 21st Century,* UNESCO, Paris.
10 Daly, H. (1996) *Beyond Growth: The Economics of Sustainable Development,* Beacon Press, Massachusetts.

Appendix II

Organizations

Centre for Action Research in Professional Practice (CARPP), School of Management, University of Bath, Bath, BA2 7AY, UK. Tel: 01225 826864. <www.bath.ac.uk/carpp/>

Center for Ecoliteracy, 2522 San Pablo Avenue, Berkeley, CA 94702, USA. Fax: 510 845 1439. <www.ecoliteracy.org>

Council for Environmental Education, 94 London Street, Reading, RG1 4SJ, UK. Tel: 0118 950 2550. <www.cee.org.uk>

Development Education Association, 29-31 Cowper Street, London, EC2A 4AT, UK. Tel: 020 7490 8108. <www.dea.org.uk>

Education Otherwise, PO Box 7420, London N9 9SG, UK. <www.education-otherwise.org>

Forum for the Future, 227a City Road, London EC1V 1JT. Tel: 020 7251 6268. <www.forumforthefuture.org.uk>

Human Scale Education, 96 Carlingcott, Nr. Bath, BA2 8AW, UK. Tel: 01275 332516. <www.hse.org.uk>

Institution of Environmental Sciences (IES), PO Box 16, Bourne, Lincs, PE10 9FB, UK. Tel: 01778 394846.

Learning for a Sustainable Future, 215 Cooper St., 2nd Floor, Ottawa, Ontario K1N 5W8, Canada. Tel: (613) 562-2238. Fax (613) 562-2244. <www.schoolnet.ca/future/>

MSc in Environmental and Development Education, Faculty of Humanities and Social Science, South Bank University, 103 Borough Road, London SE1 OAA, UK. Tel: 0207 815 5792. <www.sbu.ac.uk/fhss/eede/>

Peace Child International, The White House International Centre, 46 High Street, Buntingford, Herts SG9 9AH, UK. Tel: 0176 327 4459. <www.peacechild.org>

Schumacher College, The Old Postern, Dartington, Totnes, Devon TQ9 6EA, UK. Tel: 01803 865934.<www.gn.apc.org/schumachercollege/>

The Natural Step, 9 Imperial Square, Cheltenham, GL50 1QB, UK. Tel: 01242 262744. <www.thenaturalstep.org>

Transformative Learning Centre, OISE, University of Toronto, 252 Bloor St. West, Toronto ON M5S 1V6. Tel: (416) 923 6641 ext 2367. <www.tlcentre.org>

UNED Forum, 3 Whitehall Court, London SW1A 2EL, UK. Tel: 020 7839 1784. <www.unedforum.org> and for Earth Summit II news: <www.earthsummit2002.org>

WWF-UK, Weyside Park, Godalming, Surrey GU7 1XR, UK. Tel: 01483 426444. <www.wwflearning.co.uk> (Education site of WWF-UK dedicated to ESD).

Other useful websites

<www.esdtoolkit.org> American site from the Centre for Geography and Environmental Education, University of Wisconsin. Describes ESD, barriers to ESD, contains exercises to help schools and communities address sustainability, and includes a case study. Whole 'kit' can be downloaded.

<www.schoolnet.ca/future/> Canadian site from the foundation 'Learning for a Sustainable Future' and located on a key Canadian schools site. Includes learning outcomes, and activities for schools as well as a youth area.

<http://csf.concord.org/esf/> An American site from the 'Education for a Sustainable Future Project'. Includes curriculum resources and activities for all ages.

<www.secondnature.org> The site of 'Second Nature' which works to infuse sustainability into university teaching, research, campus operations and community outreach.

<www.unesco.org/iau/tfsd-first.html> Site of International Association of Universities' (IAU) work on higher education and sustainable development

<www.worldbank.org/depweb/english/index.htm> World Bank site on development education and sustainability. Includes learning modules, information on aspects of sustainable development, and extensive links to other areas of the World Bank site.

<www.ensi.org> Site of 'Environment and Schools Initiatives' Project in the OECD region—includes projects and initiatives by country.

<www.service-umweltbildung.de/eee/> German site on EE and ESD projects in Europe.

<www.iclei.org> Website of the International Council for Local Environmental Initiatives (ICLEI), the international environmental agency for local governments. Information on initiatives and programmes, particularly relating to Local Agenda 21.

<www.crystalwaterscollege.org.au/> Information on Sustainable Futures and Global Eco-Village Network at Crystal Waters, Queensland, Australia.

<www.unesco.org/education> For UNESCO's education programmes.

<www.tlcentre.org> Centre for transformative learning in community and adult education based at the University of Toronto.

<www.unesco.org/education/tlsf> Teaching and Learning for a Sustainable Future (TLSF) is a multimedia professional development programme prepared by UNESCO. The programme contains 25 modules for use in pre-service and in-service teacher education.

<www.ens.gu.edu.au/ciree/LSE/INDEX.HTML> Learning for a Sustainable Environment is UNESCO's guide for action-research based professional development for teacher educators in Asia and the Pacific.

<www.unep.org> Includes 'State of Global Environment' and directory of environmental education training opportunities.

<www.e4s.org.uk> British site produced by HTI (Heads, Teachers, Industry), with water and waste industries. Focus on waste management rather than broader and other aspects of sustainability.

<www.worcestershire.gov.uk/wcc/index.htm> Search under 'sustainability' and 'ESD' for information about initiatives and ESD strategy.

<www.environment.detr.gov.uk/sustainable/educpanel/index.htm> Reports and leaflets from the Sustainable Development Education Panel which advises the UK government on ESD across all sectors, in England.

<www.wri.org/enved/> World Resources Institute 'Center for Education' site.

<www.unep.org/Geo2000/pacha/index.htm> Website by young people based on their book '*Pachamama—Our Earth, Our Future*'.

<http://info.iucn.org/iucncec > Website of World Conservation Unions' Education Commission—regional reports, news, publications.

<www.ulsf.org> Site of the Association of University Leaders for a Sustainable Future. Based on the 'Talloires Declaration', and designed to help higher education institutions move towards sustainability in all areas of their operations.

SCHUMACHER BRIEFINGS

The Schumacher Briefings are carefully researched, clearly written booklets on key aspects of sustainable development, published approximately three times a year. They offer readers:

- background information and an overview of the issue concerned
- an understanding of the state of play in the UK and elsewhere
- best practice examples of relevance for the issue under discussion
- an overview of policy implications and implementation.

The first Briefings are as follows:

No 1: Transforming Economic Life: A Millennial Challenge by James Robertson, published with the New Economics Foundation. Chapters include Transforming the System; A Common Pattern; Sharing the Value of Common Resources; Money and Finance; and The Global Economy.

No 2: Creating Sustainable Cities by Herbert Girardet. Shows how cities can dramatically reduce their consumption of resources and energy, and at the same time greatly improve the quality of life of their citizens. Chapters include Urban Sustainability, Cities and their Ecological Footprint, The Metabolism of Cities, Prospects for Urban Farming, Smart Cities and Urban Best Practice.

No 3: The Ecology of Health by Robin Stott. Concerned with how environmental conditions affect the state of our health; how through new processes of participation we can regain control of what affects our health, and the kinds of policies that are needed to ensure good health for ourselves and our families.

No 4: The Ecology of Money by Richard Douthwaite. Explains why money has different effects according to its origins and purposes. Was it created to make profits for a commercial bank, or issued by government as a form of taxation? Or was it created by users themselves purely to facilitate their trade? This Briefing shows that it will be impossible to build a just and sustainable world until money creation is democratized.

No 5: Contraction & Convergence: The Global Solution to Climate Change by Aubrey Meyer. The C&C framework, which has been pioneered and advocated by the Global Commons Institute at the United Nations over the past decade, is based on the thesis of 'Equity and Survival'. It seeks to ensure future prosperity and choice by applying the global rationale of precaution, equity and efficiency in that order.

Future Briefings will deal with issues such as food and farming, globalization, local development, environmental ethics, energy policy, alternatives to genetic engineering and green technology. The Briefings are published by Green Books on behalf of the Schumacher Society. To take out a subscription, or for further details, please contact the Schumacher Society office (see page 96).

THE SCHUMACHER SOCIETY

Promoting Human-Scale Development

The Society was founded in 1978 after the death of economist and philosopher E. F. Schumacher, author of seminal books such as *Small is Beautiful, Good Work* and *A Guide for the Perplexed.* His sought to explain that the gigantism of modern economic and technological systems diminishes the well-being of individuals and communities, and the health of nature. His works has significantly influenced the thinking of our time.

The aims of the Schumacher Society are to:

• help assure that ecological issues are approached, and solutions devised, as if people matter, emphasizing appropriate scale in human affairs;

• emphasize that humanity can't do things in isolation. Long-term thinking and action, and connectedness to other life forms, are crucial;

• stress holistic values, and the importance of a profound understanding of the subtle human qualities that transcend our material existence.

At the heart of the Society's work are the Schumacher Lectures, held in Bristol every year since 1978, and now also in Liverpool and Manchester. Our distinguished speakers, from all over the world, have included Amory Lovins, Herman Daly, Petra Kelly, Jonathon Porritt, James Lovelock, Wangari Maathai, Matthew Fox, Ivan Illich, Fritjof Capra, Arne Naess, Maneka Gandhi, James Robertson and Vandana Shiva.

Tangible expressions of our efforts over the last 20 years are: the Schumacher Lectures; Resurgence Magazine; Green Books publishing house; Schumacher College at Dartington, and the Small School at Hartland, Devon. The Society, a non-profit making company, is based in Bristol and London. We receive charitable donations through the Environmental Research Association in Hartland, Devon. Schumacher Society Members receive:

• a free lecture ticket for either Bristol, Liverpool or Manchester
• the Schumacher Newsletter
• the catalogue of the Schumacher Book Service
• information about Schumacher College Courses
• a list of other members in your area, on application

The Schumacher Society, The CREATE Environment Centre,
Smeaton Road, Bristol BS1 6XN Tel/Fax: 0117 903 1081
<schumacher@gn.apc.org> <www.schumacher.org.uk>